CHINA'S DRONE AIR POWER AND REGIONAL SECURITY

DAVID SCHAEFER

Introduction by
Air Marshal **Vinod Patney** SYSM, PVSM, AVSM, VrC (Retd)

KNOWLEDGE WORLD

KW Publishers Pvt Ltd
New Delhi

in association with

Centre for Air Power Studies
New Delhi

The Centre for Air Power Studies is an independent, non-profit, academic research institution established in 2002 under a registered Trust to undertake and promote policy-related research, study and discussion on the trends and developments in defence and military issues, especially air power and the aerospace arena, for civil and military purposes. Its publications seek to expand and deepen the understanding of defence, military power, air power and aerospace issues without necessarily reflecting the views of any institution or individuals except those of the authors.

Vinod Patney
Director General
Centre for Air Power Studies
P-284, Arjan Path
Subroto Park
New Delhi 110010

Tele: (91-11) 25699131
E-mail: diroffice@aerospaceindia.org

Contents

INTRODUCTION

The outreach programme of the Centre for Air Power Studies offers non-resident fellowships to deserving candidates who are interested in writing on national security issues. Preferably, the offer is intended to encourage young scholars to research and write a monologue on a subject selected by them and approved by the Centre. The monologue is reviewed and, on acceptance as worthy of publication, is then published by the Centre as a New Delhi Paper. The paper is also uploaded on our web site. The views expressed remain those of the author and are not representative of the views of the Centre or any other organisation. The copyright of the paper rests with the author.

National security has many dimensions and, hence, the choice of a subject is indeed vast. Aspirants are required to submit a synopsis of the project and an undertaking that the work will be completed within nine months.

New Delhi Papers is a service rendered by the Centre to support the effort of young scholars, university students and those attracted to the idea of a refereed published work. The Centre will be delighted if the exercise results in a love for research and writing and the author continues with academic pursuits.

Vinod Patney
Director General
Centre for Air Power Studies

New Delhi

1. THE CHALLENGES OF DRONE DEVELOPMENT

Unmanned technology has become an essential tool for the major powers of the 21st century. Aerial drones strike a balance between other types of flying vehicles, filling a valuable niche in armed combat. Like missiles, drones are unmanned, relying on guidance controls that are operated remotely or pre-programmed in navigational instruments; but like aircraft, they are designed to be reusable and can perform multiple roles, from surveillance to strike. Without the need for a human pilot, more countries are using drones to undertake longer missions, with greater risk, and at lower cost.[1]

Since the turn of the century, the rapid innovation of this technology has introduced a novel capability for modern air power. There is growing interest in the drone programmes of other countries, and speculation about how this affects the international security landscape.[2] However, for all the interest generated by unmanned aircraft, expert discussion still largely focusses on the political and ethical questions about their use in the US-led War on Terrorism, and draws on experiences from the historical context of the last decade and a half. There is surprisingly little strategic analysis of unmanned technology as a military asset that has been slowly developing for more than a century.

This has distorted assessments of how other countries may develop and deploy drones in the future. Indeed, there are many scientific and logistical challenges for the indigenous production of unmanned aircraft. For any country, like China, which hopes to pioneer drone research in the future, it will first have to emulate the historical experience of the US military, which encountered and ultimately surmounted many problems associated with unmanned technology. As a result, this chapter traces the American experience of drone technology, as well as the more recent proliferation of drones around the world. It illustrates the factors which limit the operational deployment of drones, and the means by which they are overcome.

Early History

At first, drones were indistinguishable from missile technology. The concept of unmanned flight for military purposes can be traced back to World War I, when aviation experts began experimenting with different ways to improve their reach across the battlefield. Confronted with the deadlock of trench warfare in Western Europe, US Army engineers developed projectile technology to carry explosives over short distances.[3] Occurring at the same time as rapid progress in the science of electromagnetic signals, these early rockets were soon equipped with prototype radio control technology, allowing them to be manipulated in flight by remote operators.

In the years that followed, the first drones were developed as vehicles for target practice. Building on the earlier rockets, older aircraft were equipped with the same radio control devices for guided flight over firing ranges, but unlike missiles, these vehicles could be landed and used again. The geographic scope for unmanned flight was limited by the need to maintain radio contact within visual sight, but the US military soon adapted the technology to other tasks under the pressure of war-time. In World War II, for instance, the air force wired some converted aircraft for bombing runs, with radio operators flying in accompanying planes, controlling and directing the bombers into heavily defended targets on the ground.[4]

Specialised drone vehicles were first produced in the early years of the Cold War. The US military and intelligence community used streamlined drone vehicles to collect information in the conflict zones and territory of geo-political rivals. With improvements to inertial controls for flight stability at higher altitude, unmanned aircraft could be programmed for covert flights and then launched, relying on gyroscopes reading air pressure to navigate beyond the limits of radio contact, before being recovered at designated landing points. While details are still emerging from these years, it is known that these vehicles, equipped with automated cameras, were used for regular spy flights by the Central Intelligence Agency (CIA) to monitor strategic missile sites in China, and were floated as a possible option for photo reconnaissance of Cuba during the 1963 missile crisis.[5] Confined to reconnaissance missions over poorly defended areas, however, there was little risk of interception by enemy fighter aircraft.

This historical trajectory was altered by the Vietnam War. After high losses sustained by manned flights against North Vietnamese forces, the US Air Force turned to unmanned technology to cushion against anti-air defences. Carrying decoy weapons and sensors, drones assisted manned flights by establishing air corridors with chaff to confuse enemy weaponry, or identifying the radar frequency of missile sites that were destroyed before sorties.[6] Relying on pre-programmed navigational instruments, drones could also fly at lower altitudes to map dense jungle areas, and were sent to capture footage for damage assessment as part of the controversial US bombing campaign against North Vietnam. With more advanced equipment available for use in flight, drones eventually flew 3,435 operational sorties.[7]

In separate circumstances, unmanned air power had benefited from improved communications, self-navigation capabilities, and specialised equipment for battlefield support. However, all of these occurred under the pressure of war-time, filling niche roles that couldn't be safely attempted by manned aircraft. Following their successful contribution in Vietnam, it was predicted by aviation experts that unmanned technology could be harnessed for other roles in the US military's Cold War defence posture.[8] With the DASH helicopter drone, however, the US military soon discovered how difficult it was to build a more ambitious drone which combined all these performance attributes.

The DASH was a rotary-wing drone that was designed to hunt submarines within close proximity of aircraft carriers. Despite some promising results when it was introduced to the US Navy personnel, the aircraft slowly lost favour within defence circles. This stemmed from the plan to jointly develop the helicopter between the military branches. While designed to capture "buy in" from a range of interested stakeholders, this move actually undermined the likelihood of the technology being incorporated into the Services, as the overall concept was a consensus product that was less flexible and sensitive to the need for change. As a result, while there were several innovative uses of the helicopter by military units on deployment, there was no coordinating office which could adjust the tactical doctrine to refine its development, or initiate changes which might have secured more institutional support by making the drone more responsive to Service needs. This also meant that when the technology invariably experienced some teething issues in the harsh maritime operating environment, there was also no single voice within the US defence bureaucracy that could offer a compelling justification for continued funding. As a result, in the judgement of some experts, the programme was prematurely abandoned after a series of mediocre performance reviews.[9]

This reflected the experience of unmanned air power more generally in the later years of the Cold War. What had seemed to American military planners like pioneering, experimental technology during the Vietnam War, failed to materialise in the late 1970s and 1980s.[10] The few unmanned vehicles struggled to adapt to the more rigorous demands of high-end war-fighting in the US military's hypothetical order of battle against the Warsaw Pact, and several operational concepts fell victim to budget cuts. By the time of the Persian Gulf War, the US Navy was the only military Service to find a battlefield application for drone technology with the use of Pioneer reconnaissance drones, which were launched from the last-remaining battleship in service, to identify coastal targets for the warship's guns.[11]

This has implications for the contemporary assessment of drone technology. At various stages since the end of World War II, prototype vehicles displayed promise in operational roles that were unsuited to manned aircraft of the time. These achieved some limited successes, but did not lay the basis for a more ambitious pursuit of unmanned technology for its own sake.[12] The haphazard development of drones within the US military was due to the lack of an effective bureaucratic advocate. Without this, the technical complications and financial setbacks arising from ambitious projects undercut their potential.

The Investment Legacy

Despite this ostensible failure, there were powerful institutional advantages favouring drone research. One landmark study has pointed out that the US intelligence community's secretive "black budget" provided funding for a variety of experimental designs for drone concepts which seem infeasible even today.[13] Against the background of visible delays and errors, these various projects and vehicles slowly advanced the science of unmanned technology. Combined with an industrial policy for modernising aviation, it nurtured aviation skills and knowledge which helped to innovate and upgrade the technology over time.

This explains the breakthrough success of the Predator drone. The origin of this vehicle was a defence contract offered to a technician named Abraham Karem, who had prior experience in Israel's drone production.[14] After demonstrating a basic unmanned prototype which was simpler than the US military's crop of expensive drones, Karem was contracted to work on the "Amber" drone, a project designed to extend the surveillance reach of naval ships equipped with Harpoon cruise missiles. Building on Karem's prototype, the Amber drone's advantage was its composite mix of materials for a lightweight structural frame. This was light enough to generate vertical lift while strong enough to carry more powerful equipment on board, and permitted the installation of an advanced computer system. As a result, there was a limited number of human controllers and a more reliable flight performance. While there were teething issues with the launch device, the light weight also allowed for an easy take-off from a normal-sized runway and provided a longer time in the sky to loiter in the search for enemy objects.

By the end of the Cold War, advances in sensor and processing technology expanded the range of tasks which could be undertaken by a vehicle like Amber. In particular, satellite navigation and miniaturised electronics meant that the airborne platform was far more responsive to detailed control across a range of circumstances. Across different regions and altitudes, in any weather, variations of the Amber vehicle were capable of deep, penetrating reach into hostile territory, and could provide more lasting coverage than satellites by loitering on station.[15] With the US Air Force in need of a reconnaissance platform for use in the Balkans, the Amber design was refined, and the "Predator" was commissioned into service.[16] By the time of the North Atlantic Treaty Organisation (NATO) bombing campaign in Kosovo several years later, this drone was capable of infrared and electro-optical surveillance, laser targeting, long-distance communications relay between aircraft, and electronic jamming of Serbian air defences. Unlike manned bombers, the latest drones could assist precision-guided munitions against targets, risking anti-aircraft fire after the weather disrupted high altitude flights.[17]

It bears reminding that the initial success of the Predator was not due to off-the-shelf technology, but to the interaction between the private and public sectors. In addition to subsidising Karem's research, some technicians in the US intelligence community also developed vital enabling components, such as the data relay systems, which were incorporated into the Amber drone.[18] With the resulting vehicle frame optimised for the integration of even more components, other systems and capabilities could be added with only small refinements to the overall design. After the 2001 military invasion of Afghanistan, the American aviation industry was able to rapidly adapt to the needs of troops fighting a stabilisation campaign. The original breakthrough of the Predator in this situation opened up the way for more dedicated research and funding.[19]

This explains the notable failure of some otherwise technologically-capable countries to quickly build drones. This is particularly true of Russia, which plays a leading role in the development of aircraft, but which also experienced a widely publicised testing failure of its Stork drone in 2010.[20] This was a consequence of Soviet underinvestment from the 1980s, leaving Russian industry decades behind modern drone technology. In a similar vein, India's Defence Research and Development Organisation (DRDO) is said to be encountering problems in the development of lightweight air frames and other systems for longer flight times.[21]

Indigenous drone production requires sophisticated industry skills and knowledge of aviation design. Guidance software, miniaturised electronics, communications equipment and sensors, and propulsion systems all function together when they can be attached to a finely calibrated flying structure. Even for a national security establishment with decades of combat experience, weaving these attributes together in a reliable and capable vehicle is a demanding feat.

The Threat of Proliferation

This history adds some context to the discussion about unmanned technology in recent years. Already, by 2011, more than 70 countries were said to be in possession of vehicles which could be used for unmanned air power.[22] More important than the number of sovereign states, however, is the diverse range of organisations which can access unmanned technology, and harness it in conflict zones. With knock-off drones used by the Islamic State to record videos of its captured territory to broadcast as propaganda, this improvised technology is now available to even the weakest and least capable actors in international politics.[23]

Nonetheless, the strategic threat from drone proliferation is exaggerated. There is a wide variety of drones around the world, but very few present any serious challenge for the military security of established nation states. As one report states, "The majority of foreign UAVs that countries have acquired fall within the tactical category. Tactical UAVs primarily conduct intelligence, surveillance, and reconnaissance missions and typically have a limited operational range of at most 300 kilometres".[24] These drone models are relatively small, cheap to produce, and are equipped with only the most basic communications systems which are easily tracked. All of these introduce considerable weaknesses, limiting what they can achieve.

Consider the alleged use of North Korean drones. In April 2014, it was reported that several aircraft had crossed over the demilitarised zone separating North and South Korea, flying from the north at a low altitude to avoid detection, before hovering over several government installations. One vehicle was said to have circled South Korea's presidential mansion, Blue House, taking at least 200 photos.[25] At first, this came as a shock to the South Korean public, with the tabloid media calling for the military to strengthen border-area air surveillance systems, so that future incursions could be detected. Part of this shock related to the mode of discovery, with local civilians reporting downed vehicles, the military authorities unable to account for the initial discovery, and a widely publicised hunt for more vehicles as it became increasingly clear that North Korean forces were responsible.[26]

The incursion into South Korean air space did little harm, and the drones were severely limited in what they could achieve. The vehicles were likely reverse engineered from obsolete Chinese models, which fly along a pre-programmed route and land with the aid of a parachute.[27] This rudimentary navigation system likely explains why the vehicles failed to return to the North Korean air space. Moreover, the drones were not equipped with any weaponry, and carried basic digital cameras which stored the footage on their memory cards for retrieval after landing. At most, they are believed to be capable of flying for four hours, at no more than 20,000 ft. The only challenging feature of this incident was that the drones were so slow and small, they were either ignored or dismissed by South Korea's air radar operators.[28]

If detected, however, these tactical drones are vulnerable to interception by aircraft and air defence systems. For instance, the Shi'a Lebanese militant group Hezbollah has deployed several drones over Israeli air space in the past, using an Ayoub drone which is likely supplied by Iran to survey bases, military exercises, missile and perhaps nuclear reactor sites.[29] In a dense urban environment, where visual sighting is more likely, and which is protected by a more aggressive air defence system, the Ayoub drones are quickly identified and destroyed by Israeli fighter aircraft.[30] Tactical vehicles have virtually no means of defending themselves against other aircraft, and there are serious questions about whether their sensor equipment can withstand the gravitational force required by complex aerial manoeuvres to escape.

In these cases, the fear is not so much about what was achieved by the drone penetration, but what might have occurred if the vehicles were more advanced. While tactical drones may be upgraded and equipped with basic weapons, these will not be precision-guided missiles; and in any case, their line-of-sight communications systems are probably unable to support this type of capability. With limited space for carrying fuel on board, their range and altitude are also restricted, offering nothing more than what may be achieved by conventional ground artillery. At best, these provide brief, localised aerial surveillance which complements human intelligence of a battlefield.

This technologically is well suited to low-intensity military conflict, where casualties need to be limited. In most instances of drone warfare, the controllers are interested in providing reconnaissance and surveillance support to ground operations in a hostile area, where scouts or helicopters are likely to be met with more resistance. For instance, the Syrian regime of Bashar Assad has deployed reconnaissance drones to guide artillery strikes and troop movements near the rebel-held areas in Aleppo, Homs, and the suburbs of Damascus. According to some opposition forces, drone sightings usually precede major ground assaults; indeed, the Al Nusra branch of Al Qaeda has shot down a Yasir drone, which resembles a US tactical vehicle, allegedly supplied by Iran.[31] Similarly, UN peace-keepers in the Congo have used drones to conduct night-time surveillance flights, with the quiet engines on the vehicles able to avoid detection.[32] In both these cases, drones provided an expanded range of vision while limiting the risk to friendly forces.

To be sure, the use of tactical unmanned aerial vehicles is still of great concern to military planners. As in the North Korean example, a smaller wingspan means a diminished radar profile, allowing some tactical drones to evade detection in a cluttered air space environment, where a growing density of objects and electronic activity cloaks their movement. Unlike their larger, manned counterparts, these cheaper vehicles can also be placed at greater risk hovering near a deployed ground force, which provides some degree of surveillance as long as at least one vehicle avoids interception. With multiple, expendable vehicles being controlled simultaneously, a local drone operator is, therefore, more likely to frustrate the secure blanket of air cover which professional military forces have relied on to shield their movements.[33]

When combined with other military assets, the surveillance capabilities of unmanned aerial vehicles can be lethal. However, weaponised drones can function with accuracy only when they are equipped with a suite of sub-components, such as gyro-stabilised telescopes, laser designators, and synthetic aperture radars for penetrating bad weather. All of this adds more weight to the vehicles, and relies on in-flight stability which standard commercial vehicles struggle to deliver. Instead, more expensive and sturdy military-grade vehicles are required to properly handle an onboard system for conducting a targeted missile attack.[34] Of all the drone proliferators, Hezbollah may have come closest to innovating in this area, by attempting to convert armed Ababil drones into a rough equivalent of a cruise missile which detonates on impact. But this technology was secured through the assistance of the Iranian military, and it was reportedly unsuccessful.[35]

As a result of airborne physics, the more ambitious roles for drones are unlikely to be fulfilled by most of the world's drone proliferators. Not only is the technology more difficult to produce, but it relies on a battlefield support system.

Whereas a single unmanned vehicle which succeeds in evading detection may help to direct long-range precision fire against ground troops, this requires high-quality communications with nearby artillery or rocket forces to coordinate activity. In a contest with a modern enemy, these units must also be capable of receiving surveillance feed without being detected and attacked in turn. The sophistication of this broader enabling technology is a high threshold to breach; it is infeasible for all but a few of the world's most professional militaries to achieve on a large scale.

Instead, as the North Korean or Hezbollah examples attest, smaller tactical drones are more of a nuisance than a threat. These succeed precisely because they are primitive technology, akin to the vehicles that were used during the Vietnam War, which offer only limited improvements to situational awareness, but rarely pose a serious military threat for trained units. Indeed, vehicles with a longer range, higher altitude, more sophisticated imagery capabilities, or greater carrying capacity for weaponry, are by necessity built from sturdier wingspans to carry the weight of the added equipment.

This should caution against fear of commercial technology facilitating the spread of drones. Off-the-shelf technology cannot support the most powerful capabilities associated with unmanned air power. A series of components, and a sturdier vehicle, are required to build a strategic drone with military effect. Countries with established aviation industries, and defence aerospace skills, are more likely to capitalise on this potential.

Developmental Problems

Of the many countries experimenting with unmanned air power, only a few are experimenting with technology beyond the scope of tactical drones. And as with the US experience during the Cold War, they appear to be experiencing delays and complications. In particular, designing and producing a combat-capable drone which can respond to human control is a technological hurdle.

The need to maintain contact with human controllers imposes unique demands on the military use of unmanned aerial vehicles. Whether being remotely piloted, transmitting surveillance feed, or receiving navigational updates so that an intercepted aircraft can be salvaged, a basic requirement for drone technology is that its systems' performance can be regularly monitored while in flight. One option for secure communications is to maintain a line-of-sight radar signal, which was the technique first used by the US Air Force in World War II and the combat support missions in Vietnam, when accompanying aircraft flew near the drones equipped with radio transmitters operated by human crew. However, with the proliferation of modern anti-air weaponry, the challenge posed by electronic jamming and munitions is more pervasive than ever. Faced with a capable adversary, the need to maintain line-of-sight communications places these human controllers at risk, potentially negating the strategic value of drones.[36]

This has been mitigated by advances in communications technology, but only to a point. While more powerful equipment allows unmanned aerial vehicles to broadcast and receive signals across a wider range, it also requires a substantial increase in the carrying capacity of the vehicles. Generally speaking, the power and resilience of the data links placed on board an aircraft increase with size; indeed, an important, albeit overlooked, reason why the Predator was successfully adapted for deployment in the 1990s was that its communications did not need to be encrypted, which permitted the vehicle design to prioritise manoeuvrability over weight capacity. However, as one study points out, "When an aircraft is large, the advantages of having it unmanned are diminished, and in cases where they require a datalink to perform their mission, may even be less desirable if the security and protection of the link cannot be assured".[37]

This problem has already begun to surface. The Predator's appetite for higher-quality surveillance feed, either by supplying full-motion video or improved imagery resolution, has required the installation of relatively heavy antennas for a higher-bandwidth link with satellites.[38] In addition to this, competitive trends in the practice of electronic warfare are expected to make the technology placed aboard future drones even more sophisticated. For instance, insurgents in Iraq were discovered to be in possession of hacked video feed from US drones, which was apparently downloaded using commercial software for capturing unprotected electronic data. In response, the US military is encrypting the communications systems used by drones.[39]

While this is a problem of commercial technology available to non-state actors, professional militaries are even more capable of intercepting and disrupting communications. In a high-profile 2012 incident, the Iranian military unveiled a captured US RQ-170 "Sentinel" drone, which it claimed to have forced to the ground after interfering with its onboard Global Positioning System (GPS) by emitting a powerful spoofing signal, overriding the satellite link.[40] The exact details of this incident remain disputed, and US sources maintain that the telemetry and control functions of these drones were never compromised. Other software on the drone also provides navigation aid, to ensure that the vehicle is not wholly dependent on the GPS. Considering the fail safe systems in place, it is likely that much of the commentary on "hackable" drones was premature.[41] Even when their satellite uplink is jammed, US military drones can be expected to revert to other navigational software, and return to base.

Electronic interference is not a problem for unmanned aerial vehicles alone. The general thrust of fifth-generation aircraft in the global aviation industry is towards networked operations, which rely extensively on secure communications and data transfers across multiple systems simultaneously. But unlike manned aircraft, the technical challenge for unmanned air power is more likely to encourage a greater overhaul of vehicle design: a large part of the success of drone vehicles is due to their light weight, so while efforts can be made to shore up communications against electronic interception, the added weight that results from this innovation also introduces further design trade-offs. For instance, a heavier vehicle will burn through more fuel in flight, reducing loitering time above targets; it will almost certainly be less versatile in flight, as the internal mechanics of current equipment attached to the vehicle are more sensitive to sharp movement.[42]

This also raises doubts about the digital capacity for communications. Improving the reliability of onboard systems to relay secure communications eats through more bandwidth than unencrypted transmissions. Already, however, the volume of data which results from drone operations is staggeringly large: in 2009 alone, it was reported that US unmanned aerial vehicles across the world had produced a cumulative 24 years of video to analyse.[43] At least for the US, with a number of peace-time intelligence gathering operations at any one time, the vast amount of data from existing drone missions is placing greater strain on digital infrastructure. For example, one Global Hawk drone needs 500 Mbps of bandwidth, which is five times the total amount of bandwidth that is estimated to have been used by the entire US military during the Gulf War.[44]

For the US military, the communications challenge has also led to the development of an elaborate support network to facilitate the global reach of operations. In the case of the bombing campaigns in Yemen, Iraq, and Afghanistan/Pakistan, a single Reaper or Predator drone relies on a much larger contingent of ground personnel than manned aircraft. Depending on the aircraft, these have been cited as 168 people (Predator), 180 (Reaper), and 300 (Global Hawk).[45] Many of these people are stationed at one of several regional bases for operations which provide basing, fuelling, maintenance, and repair facilities for the vehicle in its theatre. With a glut of data from surveillance flights, there is a corresponding need to train and equip an analytical workforce which processes the vast amount of collected information.[46]

In the absence of local sites within the conflict zone, dedicated space assets are also needed to support the expanded operating range of drone flights. Since 2001, the US military has relied extensively on military and commercial satellites to transmit real-time data from aircraft in Iraq and Afghanistan to off-site controllers. According to space experts, the deepening reliance on information systems is placing strain on digital communications, even as more satellites have been leased to free up overall capacity; indeed, the massive growth of data used by US military forces deployed around the world has caused some drone operations in Iraq and Afghanistan to be cancelled for want of more bandwidth.[47] This has not only limited flying hours, but also affected other capabilities: one director from Northrop Grumman conceded that many sensors on the US-made Global Hawk surveillance drone would operate at a much higher tempo if the data links were capable of supporting it.[48] With more improvements to surveillance equipment on unmanned vehicles in the future, this burden is only expected to increase for the US military over time.

In the light of the communications problem, the trends in research and development for unmanned air power point towards larger models of drone aircraft over time. Competitive vehicles need to be heavier, carry a greater range of equipment, and rely on a network of ground and space assets and a highly skilled workforce to make use of the information. All of these demands will be imposed on drone proliferators in the future, and place a premium on reliable space-based infrastructure for data transmission.

The Strategic Context

Until recently, the discussion of unmanned air power has largely neglected these limitations. The appearance of armed drones was believed to usher in a "new era of warfare", with unpiloted vehicles capable of extraordinary reach and penetration. The cheap cost, diminished risk for the pilot, and absence of a cockpit has led some commentators to question whether the technology would ultimately supplant manned aircraft.[49]

The reality is that this development took place in a favourable strategic context. For the US military, the lack of communications security and vulnerability to anti-air artillery did not pose much of a challenge for drones operating in Afghanistan and Iraq. Traversing combat theatres with guaranteed air control, faced with insurgents who lacked sophisticated weapons, slow-moving drones could stay above the fray for a long time without needing to undertake demanding manoeuvres. Their limited capacity to field missiles and sensors provided sufficiently targeted capabilities to be of use for intelligence-led, counter-insurgency campaigns.

On those few occasions when drones were placed outside this context, however, they struggled against competing technology. Before the US invasion of Iraq, for instance, a Predator drone was shot down by an Iraqi MiG-25 fighter after its controllers tried to engage the aircraft over a no-fly zone near the Persian Gulf.[50] Similarly, the deployment of unarmed Predators in the NATO campaign in Kosovo, where the vehicle's laser designators were used for targeting guidance, witnessed substantial losses of drones after poor weather conditions required them to operate closer to Serbia's air defences on the ground.[51]

For more than a decade, the US enjoyed a permissive air environment, which was conducive to more experimentation. In particular, the covert bombing campaign against terrorist havens in countries like Pakistan, Yemen, and Somalia was facilitated by the policy of host governments, which encouraged and assisted with local intelligence for targeting decisions. As a result, hastily produced models were upgraded over time, with improvements to sensors and weaponry, even as they delivered a valuable strategic effect for the US as a prototype. While the Pakistan Army formally complains about violations of its air space, there is little doubt that it is capable of shooting down foreign drones if it wanted to. This compliance is not assured for other countries which might want to deploy unmanned air power in the future.

But while these drones were adapted over time, they still represented a choice for a specific capability. The Predator and Reaper classes of strategic drones were designed to fill an operational role, which involved trade-offs between performance features. They are slow-moving, which offers persistence; they rely on a heavy-bandwidth data link to transmit imagery in real-time; they are built on engines which produce enough vertical lift to carry some limited equipment over great range; and they do not cost as much as manned aircraft. This combination of attributes is suited to operations in Iraq and Afghanistan, but it sacrifices the speed, manoeuvrability, and stealth that could be useful in other roles. This is in contrast to rotary-wing drones used in naval support roles, for instance, which carry larger payloads, and are more versatile and reliable, but are slower and have higher up-front costs.

As a result, with deployments shifting, the direction of future drone technology for the US military is uncertain. Outside the conflict zones in Afghanistan and Iraq, which have shaped drone procurement, the navy is researching an armed drone that can be launched from an aircraft carrier for likely deployment in the Asia-Pacific. Despite this ambition, there is uncertainty about whether the navy wants a more affordable vehicle which provides intelligence support to other aircraft, or a long-range bomber which can penetrate enemy air defences. These rival mission roles have led to a protracted debate about which capabilities to embrace, and which to forego.[52] Unlike in Afghanistan or Iraq, the existence of a hypothetical Chinese enemy with sophisticated air defence systems will make it more difficult to resolve with incremental changes to existing models.

The Limits of Physics

As other countries explore options for armed drones, they will confront more hostile scenarios for air operations than experienced by the US military. This will require them to address the weaknesses and improve the capabilities of drones across several performance areas, which will make the vehicle designs more complex in the process. This raises the bar for acquisition of the latest, costly armed drones. While tactical drones may be widely available, many countries will be priced out of the market for new strategic vehicles.

The cheaper price of unmanned technology has long been one of the key advantages for drones compared with manned aviation. This is particularly because modern aircraft have experienced an inexorable rise in per-unit cost over the last several decades, as the task of integrating sophisticated inputs becomes more technically demanding for a streamlined airframe.[53] Compared with the spiralling cost of procurement for the latest types of fighter and bomber aircraft, drones have been touted as an alternative to keep air power financially sustainable over the long term.[54]

On the face of it, the procurement of Reaper and Predator drones over the last decade has been cheaper for the US military than manned aircraft which offer a roughly equivalent capability. This makes sense because manned aircraft include a cockpit with a series of integrated life support systems, which require structural and weight changes that need to be offset with a more powerful engine system. These mean that any manned aircraft has greater demands placed on it than an unmanned aerial vehicle which is designed for a cheap and efficient working role. As one study argues, "the cheapest surveillance and armed drones will always be less costly than the cheapest surveillance and strike aircraft".[55]

But there is also a reason to be sceptical. For one, the budgeting of drones underreports the enabling capacities which are critical for the vehicles to function, such as the remote basing and computer terminals for controllers. While infrastructure support is also required for manned aircraft, these are almost always interoperable with a greater range of platforms, allowing the costs to be shared in a way which does not apply to drones. There is also the case for a more considered estimation of the full life-cycle costs of drone aircraft, as unmanned control is vastly more liable to failure arising from pilot error and software failure, requiring higher replacement and maintenance costs. Indeed, one analyst has estimated that the annual operating costs of the Reaper drone is roughly four times that of an F-16 or A-10 aircraft.[56] With all these complicating elements included in the budget projections, the picture looks much less clear.

More importantly, further development of technology will introduce new design complexity, and this can be expected to raise procurement costs over time. Because manned aircraft are inherently more flexible, they can fulfil a greater number of roles, from surveillance to ground support to strategic strike, with adequate performance across a range of areas. This is not so much the case with unmanned technology; instead, with new technologies being explored, the many different design features and advantages to choose from are working at cross-purposes.

Take, for instance, the next type of munitions used by Reaper drones. Given that the advantage of these aircraft is in loitering for long periods of time, there is a strong case to be made to manufacture smaller warheads in their munitions, so that the low mass permits the vehicles to operate for longer without burning as much fuel to stay aloft.[57]But this sacrifices payload capacity over endurance, and an alternative might involve designing a larger weapon, which would entail some reduction in the imagery sensors to make room for the extra weight.

This logic applies to all complex military systems, from submarines to armoured vehicles. It also casts doubt over the technical feasibility of an armed combat drone, which is a concept that has received greater interest in recent years. In order to defeat potential adversaries in the air, a tactical fighter needs to be capable of performing sharp turns and versatile manoeuvres. This requires a sturdy wing design that can withstand high gravitation forces, but the composite materials on the Predator are not believed to be seriously capable of fulfilling this demand.[58] Instead, a more resilient structure will increase the weight of the vehicle, and this, in turn, will require a more powerful engine, as well as more fuel or battery power if it is to be capable of flying a similar distance. Taken together, the final product is, therefore, likely to involve a much heavier design, with a much higher cost.[59]

In addition, the resilience of technology onboard many US drones has already struggled under the pressure of combat. The reliability of the software and equipment on the Predator has been a consistent irritation for the US military, because many new parts in the operating system were rapidly integrated into the existing models after 2001, with more than 400 crash incidents occurring since that time.[60] As with the first generation of military aviation at the beginning of the 20th century, military drones have also proven difficult. Issues like the impact of adverse weather conditions, inconsistent engine power, and software glitches have all regularly cropped up.[61]

Some of this can be explained as a natural teething process for the incorporation of new technology. However, the sensitive nature of unmanned equipment is likely to become more of a problem as a drone's flying ability improves. The lightweight communications systems or electronic sensors could be easily disrupted or damaged by high-end aerial manoeuvres in tactical combat. It may be that more advanced or expensive technology could be used as an alternative in the future, so that it could withstand the pressure; but this might also divert more power from the drone's electrical system, with the risk of limiting the processing power of the computer. In order to conduct tactical manoeuvres, however, a combat drone will need an optimised processing system to perform more cognitive tasks autonomously. This is not to say that armed combat drones are impossible, but that satisfying the many demands of tactical flight will require a larger, sturdier vehicle with more available power than is currently feasible.

At the very least, achieving this will be vastly more expensive, and this brings with it added risk for project development. In the late 1970s and early 1980s, US defence planners explored a similarly ambitious drone idea for air support to ground troops called "Aquila". With a number of possible improvements to the prototype floated by designers, the Aquila began to resemble a compromise between competing mission roles, with multiple requirements placed on the vehicle. This led to rising unit costs, and continued delays as further design complexity was required to integrate the growing number of capabilities onto a single vehicle, until political support withered away and the programme was cancelled in 1983.[62]

The fate of the Aquila is the danger which confronts the development of armed combat drones in the future. As new models balance innovations in each area of aircraft performance, the general trend is towards vehicles which can supply multiple features and perform a wider range of tasks. This will inevitably drive up the financial cost of production and without an overarching vision for exploiting unproven technology, the many remaining opportunities for battlefield application and scientific investment may not be exploited. Unless a country's military acquisitions process is prepared for delays and rising costs, the next generation of armed, competitive drones is unlikely to be achieved.

As a result, a country like China, with plans to harness drone technology for air power, needs to meet a minimum set of preconditions. Beyond purchasing a copy off-the-shelf, strategic-level drones are not easily acquired. Instead, the history of the American experience with drone technology points to several major hurdles which need to be scaled for sustainable production and operation. These are:

- A national aerospace industry with the skills for aircraft design and integrating components.
- A military which promotes experimentation and is willing to incur sustained financial costs.
- A communications infrastructure with space platforms for the real-time transmission of data.
- An operating environment which requires persistent use of drones for training and development.

The absence of these factors explains why some countries have experienced problems with a technology that appears so widely available and easy to use. This is doubly so for countries which hope to develop armed combat drones in the future. The question is whether China has the scientific and industrial resources to surmount this challenge.

Notes

1. A good summary of UAV characteristics can be found in Lynn Davis et. al., *Armed and Dangerous? UAVs and US Security*, RAND, May 2014 Research Paper, p.1.
2. Kelvin Wong, "Armed Drones in Asia", RSIS Commentary, no.97, August 23, 2010.
3. For a basic outline of this story, see Jeffrey Sullivan, "Evolution or Revolution? Rise of UAVs", *Technology and Society*, vol.25, no.3, Fall 2006, pp.43-49.
4. Andrew Callam, "Drone Wars: Armed Unmanned Aerial Vehicles", *International Affairs Review*, vol.18, no.3, p.18.
5. Thomas Erhard, *Air Force UAVs: The Secret History*, Mitchell Institute for Airpower Studies, July 2010 Research Paper, pp.7-11.

6. "RPVs to Play Electronic Warfare Role", *Aviation Week and Space Technology*, January 22, 1973, p.59.

7. Sullivan, n.3.

8. Barry Miller, "Remotely Piloted Aircraft Studied", *Aviation Week and Space Technology*, June 1, 1970.

9. See, for example, Benjamin Armstrong, "Unmanned Naval Warfare: Retrospect and Prospect", *Armed Forces Journal*, December 20, 2014.

10. William Broad, "The US Flight from Pilotless Planes", *Science*, July 10, 1981, p.188.

11. Peter Singer, *Wired for War: The Robotics Revolution and Conflict in the 21ˢᵗ Century*, 2009, pp.56-57.

12. See, for example, Tony Capaccio, "New Failure May End Tacit Rainbow's Misery", *Defense Week*, September 24, 1990, p.7.

13. Erhard, n.5, pp.31-32.

14. See for example "The Dronefather", *Economist*, December 1, 2012.

15. David Fulgham, "UAVs Pressed into Action to Fill Intelligence Void", *Aviation Week and Space Technology*, August 19, 1991, p.59.

16. John Boatman, "Tier 2 Plus: Taking the UAVs to New Heights", *Jane's Defence Weekly*, August 12, 1995, p.34.

17. Richard Whittle, "Predator's Big Safari", Mitchell Institute for Airpower Studies, August 2011 Research Paper, pp.131-135.

18. Frank Stickland, "The Early Evolution of the Predator Drone", *Studies in Intelligence*, vol.57, no.1, March 2013, pp.4-5.

19. John Tirpak, "The RPA Boom", *Air Force Magazine*, vol.93, August 2010.

20. David Axe, "Russian Drones Lag U.S. Models by 20 Years", *Wired* online, August 6, 2012, accessed at: http://www.wired.com/2012/08/russian-drones/

21. Aditi Malhotra and Rammohan Viswesh, "Taking to the Skies: China and India's Quest for UAVs", *Journal of Indian Ocean*, 2014, p.7.

22. US Government Accountability Office, "Nonproliferation: Agencies Could Improve Information Sharing and End-Use Monitoring on UAV Exports", July 2012 Report, p.10.

23. Peter Bergen and Emily Schneider, "Now ISIS has Drones?", *CNN* online, August 25, 2014, accessed at, http://edition.cnn.com/2014/08/24/opinion/bergen-schneider-drones-isis/

24. n.22, p.11.

25 William Gallo, "Alleged North Korean Drones Fly Over Seoul Presidential House", *Voice of America*, April 3, 2014.

26. See, for example, "S. Korea Finds Third Suspected N. Korean Drone", *Yonhap*, April 6, 2014.

27. Sang Jun Lee, "North Korean UAVs Likely of Chinese Origin", Centre for Strategic and International Studies, *cogitASIA* online blog, April 21, 2014, accessed at http://cogitasia.com/north-korean-uavs-likely-of-chinese-origin-2/

28. Van Jackson, "Kim Jong Un's Tin Can Air Force", *Foreign Policy* online, November 12, 2014, accessed at http://www.foreignpolicy.com/articles/2014/11/11/north_korea_drones_kim_jong_un_seoul_tin_can_bomb_deadly

29. "Hezbollah Drone Photographed Secret IDF Bases", *Jerusalem Post*, October 14, 2012.

30. Isabel Kershner, "Israel Shoots Down Drone Possibly Sent By Hezbollah", *New York Times*, April 25, 2013.

31. Wim Zwijnenburg, "Drone-tocracy? Mapping the Proliferation of Unmanned Systems", Oxford Research Group, *Sustainable Security* blog, October 8, 2014, accessed at http://sustainablesecurity.org/2014/10/08/drone-tocracy-mapping-the-proliferation-of-unmanned-systems/

32. Somini Sengupta, "Unarmed Drones Aid UN Peacekeeping Missions in Africa", *New York Times*, July 2, 2014.

33. Quoted in D.J. Hurt, "Fires in Decisive Action: Developing Capabilities Required to Win the Next Phase 2", *Fires Magazine*, July-August 2012, p.37.

34. Davis et.al., n.1, p.3.

35. Ronen Bergman, "Hezbollah Stockpiling Drones in Anticipation of Israeli Strike", *al-Monitor* online, May 15, 2013, accessed at http://www.al-monitor.com/pulse/tr/security/01/05/the-drone-threat.html

36. Courtney Howard, "UAV Command, Control, and Communications", *Military & Aerospace Electronics*, vol.24, no.7, July 11, 2013.

37. Davis, et.al., n.1, p.5.

38. Ibid., p.2.

39. Ellen Nakashima, "Encryption of Drone Feeds Won't Finish Until 2014, Air Force says", *Washington Post*, December 19, 2009.

40. David Cenciotti, "Iran Seizes a US Stealth Drone By Taking Over Controls. Maybe", *The Aviationist*, December 4, 2011, accessed at http://theaviationist.com/2011/12/04/iran-drone/

41. David Axe, "Iran Probably Didn't Hack CIA's Stealth Drone", *Wired* online, April 24, 2012, accessed at http://www.wired.com/2012/04/iran-drone-hack/

42. Winslow Wheeler, "The Reaper Revolution: The MQ-9's Cost and Performance", *TIME* online, February 28, 2012, accessed at http://nation.time.com/2012/02/28/2-the-mq-9s-cost-and-performance/

43. Christopher Drew, "Military is Awash in Data from Drones", *New York Times*, January 10, 2010.

44. Jeremiah Gertler, "US Unmanned Aerial Systems", Congressional Research Service, January 2012 Report, p.17; see also www.globalsecurity.org/space/systems/bandwidth.htm

45. Shashank Joshi and Aaron Stein, "Emerging Drone Nations", *Survival*, vol.55, no.5, 2013, p.56.

46. Ibid., p.62.

47. Brett Biddington, "Skin in the Game: Realising Australia's National Interests in Space to 2025", Kokoda Foundation, May 2008, Research Paper no.7, p.39.

48. Grace Jean, "Remotely Piloted Aircraft Fuel Demand for Satellite Bandwidth", *National Defense*, July 2011.

49. Paul McLeary, "Former ISR Chief Calls for More Autonomy in UAVs", *Aviation Week*, January 27, 2012.

50. Brian Palmer, "Is It Hard to Kill a Drone?", *Slate*, June 6, 2012.

51. Ibid.

52. Marina Malenic and Zachary Fryer-Briggs, "Budget Concerns Hold Up UCLASS RfP Public Release", *IHS Jane's Defence Weekly*, September 19, 2014.

53. Mark Arena et. al., *Why Has the Cost of Fixed-Wing Aircraft Risen?* RAND, 2008 Research Monograph, pp.79-80.

54. See, for example, "The Last Manned Fighter", *Economist*, July 14, 2011.

55. Joshi and Stein, n.45, p.56.

56. Winslow Wheeler, "The Reaper Revolution: Revolutionary…or Routine?", *TIME* online, March 2, 2012, accessed at http://nation.time.com/2012/03/02/5-revolutionary-or-routine/

57. David Hayes, "The Weaponisation of Future Unmanned Aerial Vehicles", *RUSI Defence Systems*, Autumn/Winter 2012, p.79.

58. See Wheeler, n.56.

59. Prem Mahadevan, "The Military Utility of Drones", CSS Zurich Analysis, no.78, July 2010, pp.2-3.

60. Craig Whitlock, "When Drones Fall From the Sky", *Washington Post*, June 20, 2014.

61. Winslow Wheeler, "The Reaper Revolution: Finding the Right Targets", *TIME* online, February 29, 2012, accessed at http://nation.time.com/2012/02/29/3-finding-the-right-targets/

62. On this point, see Bruce Berkowitz, "Sea Power in the Robotic Age", *Issues in Science and Technology*, vol.30, no.2, Winter 2014.

2. China's Emerging Drone Infrastructure

The previous chapter provided a brief history of the US-led drone development, highlighting several underlying challenges which are often neglected in the academic treatment of drone warfare. This chapter surveys these issues—industrial, organisational, communications and strategic—in the context of China's own military modernisation.

Unlike most other countries interested in unmanned technology, China's military and political leaders have launched an ambitious programme for the indigenous production of drones, and possess the tools to convert this technology into an effective military capability. For several decades, the People's Liberation Army (PLA) has already been experimenting with basic vehicles, and recent years have witnessed several aircraft designs and operational concepts which replicate, and in some cases, surpass, the armed drones being developed by the US military.

Without much accompanying detail to scrutinise for problems, this has given rise to concern. In one widely-quoted example, a 2012 report by the US Defence Science Board found that the "military significance of China's move into unmanned systems is alarming".[1] Nonetheless, serious questions remain about the training and operational readiness of the PLA's drones, as well as its capacity for innovation. Indeed, the image of rapid progress in unmanned technology is almost certainly exaggerated for political effect.

Industry Skills and R&D

China has a skilled workforce and economic infrastructure to support and expand drone research and development. Since the 1990s, the country's aviation industry has been scaling the entry barriers for modern commercial and military operations, relying on industrial policy for import substitution. This provides a growing knowledge base for modern aircraft development, and will flow into the production of unmanned technology.

There are signs of an aggressive development strategy for China's aviation industry which are common to other areas of investment by the country to secure military advantage. By opening up its domestic labour force to foreign investment, and deepening its involvement in the global supply chains of aerospace companies, China has been slowly accumulating expertise in dual-use technology for air power.[2] The country's large domestic market has been used as leverage in negotiations over joint development projects, as foreign firms are generally required to deal with local sub-contractors, securing some degree of intellectual property transfer as well as skills in design and production. At present, this typically involves the use of older equipment, and valuable components assembled in China are mostly still made abroad, limiting the potential for industrial leakage. The experience of incorporating this technology provides some exposure to machining techniques, but it has not yet closed the gap between China and the world's most advanced technology.[3]

However, this has been matched with more dedicated resources and organisational support for local production. Chinese industrial development in aviation is coordinated by the Aviation Industry Corporation of China (AVIC), a state-owned consortium for aerospace construction and defence operations. This acts as an umbrella organisation for a number of projects, with the aim of sidelining foreign penetration of the Chinese market over time. For instance, the C919 project managed by the Commercial Aircraft Corporation of China (COMAC), a subsidiary of AVIC, illustrates the determination of the Chinese state to foster aerospace capabilities at the expense of competitors. The C919 is a narrow-body passenger aircraft, part of a commercial bid to capture a large share of the global aviation market, which is currently dominated by Boeing and Airbus vehicles. Commercial airlines in China are being pressured to buy greater numbers of the C919, despite the technical problems and delays associated with its roll-out.[4] Vast resources are being funnelled into the project to avoid too much delay; indeed, local engineers working on the aircraft are said to be receiving twice the standard industry wage.[5]

China is willing to incur significant costs to achieve this goal. Even if a timely schedule is maintained by COMAC, the life-cycle economics of heavy passenger aircraft like the C919 stretches over several decades, and includes higher than usual maintenance and fuel costs. This will bring more risk for its commercial buyers than is implied in the shelf price, making it even more uneconomical to purchase in such large numbers. And by the time COMAC is able to begin fulfilling its contract sales, the C919 will likely be outdated compared to the latest models offered by rival foreign firms. This is an example of how China's political and military leaders are willing to absorb huge inefficiencies and sub-optimal allocation of resources to claim the technological edge. The motive is not profitability in the short term, but to position China's manufacturing workforce for the production of more sophisticated models in the years ahead.

This is a sign of China's massive industrial capacity for research and development of air power, which is now extending to drone technology. Major subsidiaries of AVIC are clearly being directed towards more indigenous drone production, including the Chengdu Aircraft Industry Group, the Guizhou Aircraft Industry Corporation, and the Shenyang Aircraft Company.[6] Indeed, according to one study, the Guizhou Aircraft Industry Corporation "is expected to become a full service manufacturing, testing and service 'base' for the PLA's UAVs", and will be operational by 2015.[7] While these organisations are the few which offer public information about their research agendas, it is believed that "every major arms manufacturer in China now has a devoted drone research center".[8]

Military Organisation and Planning

While China's industrial promotion secures dual use civilian technology, there is a centrally directed ecosystem designed to channel any scientific gains for military innovation. Overseen by Chinese defence authorities, this nationwide effort spans commercial organisations, research centres, and university laboratories.[9] Each of these has a particular, if sometimes overlapping, role to play in the development and exploitation of drone technology.

Traditionally, aviation research has received targeted funding through project offices in the Chinese Ministry of Science and Technology. A number of government-sponsored grants have been offered to researchers in unmanned technology, advancing development for military purposes.[10] For instance, the "863 program" within the ministry provides funding for the Beijing University of Aeronautics and Astronautics; and this university has pioneered unmanned technology research in China, including drone models which have been displayed at the Zhuhai Air Show.[11] In addition to this, the "973 program" in the same ministry, responsible for multi-disciplinary "cutting edge" technology, has also provided funding to the Nanjing University for Aeronautics and Astronautics, which has used this to publish academic papers on drone-related tasks, such as using computer science to land vehicles on ships.[12]

These efforts also receive support from Chinese military personnel allegedly involved in cyber espionage. Computer security experts in the Western nations have detected a persistent campaign of intellectual property theft over the last decade, and have traced this to the location and computers of units in the PLA. It was recently reported that a number of American companies experimenting with drone technology have been singled out for targeted cyber-attacks, and this undoubtedly accesses at least some technical details to speed up the modernisation of China's drone capabilities.[13]

Within the PLA, there are several clusters of drone activity. The basic division of labour is between the General Staff Department (GSD), which focusses on joint-force mission planning, to integrate unmanned systems into doctrine; and the General Armaments Department (GAD), which focusses on research and development, as well as advising the Central Military Commission (CMC) on resource allocation and industrial projects, which provides guidance on AVIC's major projects. GAD's national test lab for drones is housed in the Northwestern Polytechnic University; this institute is said to have produced more than 40 variations of drones in the past, and owns 90 per cent of the Chinese domestic market for unmanned vehicles.[14] Within the GSD, a number of sub-departmental branches exercise influence over drone activity, by providing specialised advice on certain operational requirements. For instance, within the GSD intelligence department, the S&T equipment bureau, 55th research institute, and intelligence reconnaissance bureau, each has an unmanned component to its area of responsibility.[15] The same is true of the electronics counter-measures and radar departments, which are involved in the development of drone systems.[16]

Each major Service in the PLA – army, navy, and air force – also includes equipment and intelligence departments exploring drone operations. Most of these also operate smaller, tactical vehicles for battlefield support and training, including an air force unit at Fuzhou. By contrast, the Second Artillery Headquarters, which is responsible for China's missile and rocket forces, maintains a battalion that conducts drone operations, but this is very likely to be developing more strategic-level vehicles for military contingencies involving missile strikes. Similarly, the GSD's intelligence department maintains a brigade or regimental-level unit which likely operates drones for sensitive intelligence gathering and reconnaissance tasks.[17]

The result of all this is a vast apparatus for implementing and innovating drone technology.[18] Backed by industry skills and political will power, the PLA can experiment with drones in a wide variety of circumstances and operational roles, with the eventual goal of scaling the technology deficit between China and potential rivals like the United States.

Space and Digital Communications

Ever since China decided to develop nuclear weapons in the early years of the Cold War, efforts to master the science of missile launch and remote tracking systems have also nurtured an expertise in space. These developments have reached the stage where communications and navigation links can enable the operational deployment of drones.

Having undertaken aerospace research over the last several decades, China now possesses an advanced space flight industry with interconnected manufacturing hubs, launch sites for different orbital ranges, dedicated command and control facilities, and integrated ground stations and surveillance ships for tracking space activity. Since opening up to commercial activity in the 1980s, China's space sector has enjoyed a sustained growth in the number of shuttle launches, which overtook the US on an annual basis for the first time in 2011.[19] In a sign of the industry's growing importance under the leadership of President Xi Jinping, several directors of Chinese aerospace conglomerates were recently promoted within the political ranks of the Communist Party. This points to a growing presence in, and mastery of, space operations in the future.

This is significant because the challenge of relaying processing information and sensor data between the drone machine and human controller is likely to be a major hurdle for many other countries with drones. One study of drone proliferation argues, "…in most cases emerging drone nations will use the most capable of these platforms in smaller numbers and in more local contexts than is sometimes assumed".[20] Faced with this challenge, satellites offer a combination of persistence, range, access, and accuracy that is highly valued for surveillance and communications. There is a shortage of transmission capacity in the commercial satellite market which can be leased for coverage in other parts of the world; in China's case, however, the same constraint does not apply.

Instead, China has a range of space platforms which can enable unmanned flight beyond its territory. Most notable is the indigenous navigation system, Beidou ("Compass"), which has been developed since 2000. Beginning as a regional service, Beidou is now expanding to provide coverage for users located anywhere in the world, and by 2020, the system will rely on a constellation of 35 satellites capable of transmitting data for navigation.[21] While this brings commercial benefits, the primary motive behind the creation of Beidou is almost certainly the People's Liberation Army's desire to reduce its dependency on the US-managed GPS. Equipped with Beidou, Chinese military assets, including unmanned aerial vehicles with onboard receivers, are able to locate their position to within several metres of accuracy.

China also boasts of a range of communications and reconnaissance satellites, which help to remotely operate military assets beyond its territory through imagery and electronic sensors. While the technical details of some onboard systems are not publicised, new satellite models like the Yaogan series, which are used for scientific missions, are also widely believed to contain a balanced range of remote sensing equipment that can be harnessed for military purposes. At the very least, China has electro-optical and synthetic aperture systems which provide imagery of the country's neighbourhood, and sensors capable of detecting electromagnetic activity.[22]

As a result, China now possesses a ready-made network to control drones beyond its territorial limits, and to do so with improved accuracy and reliability. Dedicated space platforms can assist drones in mission planning, cue targeting, provide damage assessment, offer reference points for navigation, and maintain regular contact between onboard sensors and software equipment with human controllers. This is particularly valuable for a maritime operating environment, where local surveillance units, such as ground observers or embedded intelligence networks, are not as readily available for targeting guidance and battlefield assessment.

The Growth of the Drone Fleet

All of this economic, military, and space-based progress lays the groundwork for China to field a fully-capable drone fleet in its strategic neighbourhood. This is reflected in the country's military inventory, as the PLA has been supplying the latest prototype vehicles to field units.

To be sure, China's interest in military drones stretches back several decades. The first unmanned aerial vehicles are known to have been reverse-engineering from technology acquired through military partnerships in the early years of the Cold War. Beginning with target drones supplied by the Soviet Union in the 1950s, and later augmented with the transfer of US drones captured by North Vietnamese fighters during the Vietnam War in the 1960s, China was able to manufacture basic vehicles for low-altitude flying tasks like training and fire drills. The Wu-Zhen 5, was a spinoff from the American AQM-34 Frisbee drone; while the Chang-Kong 1 was built from the Soviet Lavochkin La-17, and used to monitor the radiation fallout from China's nuclear tests at close range.[23]

But after the Sino-Soviet split in the 1960s, which marked a period of greater isolation for Chinese foreign policy, the country's research institutes were mostly confined to innovating these models. This has produced dozens of reconnaissance and targeting drones, including the Chang-Hong 1, a variation of the US drone used in the Vietnam War. Beginning in the 1970s, the People's Liberation Army Air Force (PLAAF) also converted some of its older aircraft, including the J-5, J-6, and J-7 fighters, into remotely piloted vehicles.[24] In the 1990s, an unknown number of Harpy drones, which are designed to detect and attack radar installations, were also purchased from the Israeli military. In 2005, after China contracted with Israel to upgrade this fleet, with what some experts suggested was sensor technology for detecting radar sites that were switched off, Israel pulled out of the arrangement under American diplomatic pressure.[25]

Since the turn of the century, however, China's efforts at indigenous production have dramatically expanded. According to one study, "over the course of the 10th and 11th Five-Year Plans (2001-10), the PLA has apparently committed to investing in a world class unmanned systems capability".[26] This has quickly borne results: over the last decade, the drone technology displayed at China's biennial air show has evolved from art drawings to operational models. Vehicles are now prominently displayed in parades and at industry gatherings, take part in military training exercises, and are reportedly incorporated into some PLA units.

Recent estimates from Taiwanese sources have placed China's military drone inventory at over 280 vehicles by mid-2011, although this number has undoubtedly grown in the interval.[27] It seems likely that this number is mostly composed of tactical, unarmed drones (both fixed and rotary wing). Many of these vehicles are variations of the ASN-200 series drones, a vehicle originally developed in the 1990s, and now said to be enjoying widespread use by the People's Liberation Army and Navy in reconnaissance roles.[28] These vehicles are necessarily limited in range and payload capacity, but it is notable that they are equipped with data links to ground stations, which allow for real-time transmission of information.[29] Along with other tactical drones, like rotary-wing drones, they form part of a modernising fleet for joint-force operations for smooth intelligence gathering and situational awareness. In fact, as a US Congressional review noted, "[b]ecause of their applicability to a wide range of military missions, China will likely continue to focus on the development of tactical-level UAVs, particularly those that operate at low-to-medium altitude and have close-to-medium range".[30]

By contrast, more ambitious concepts for drone aircraft, including strategic vehicles which closely mirror the latest US military technology, have recently emerged. While official information is limited, and media coverage provides only rudimentary details, a large amount of raw data, including pictures and gossip from aviation enthusiasts, has surfaced online. Taken together, this points to several models under development, with some likely finalised for large production runs.

Projects under Development

The Wing Loong or "Pterodactyl" is a medium altitude, long range drone. This vehicle is a rough equivalent to the US-made Predator drone. A joint project between Hongdu Aviation Industry Group and Shenyang Aviation Corporation, the Wing Loong began development in 2005, conducting its first test flight in 2007, and was publicly unveiled in 2012. It can fly for 20 hours, up to a range of 4,000 km, at a maximum speed of 175 mph, and carries a payload of 200 kg.[31] The Wing Loong is believed to be able to mount two missiles, and has been displayed next to several types of armaments. The most likely candidate is the BA-7 semi-active, laser guided air-to-ground missile, weighing 47 kg with a range of 7 km; there is also a LS-6 50 kg miniature guided bomb; the YZ-102A precision-guided missile and the YZ-121 laser guided missile.[32] These are almost certainly guided by infrared sensors, as local officials using the Wing Loong for internal security have talked about its value in locating targets in the dark of evening.[33]

The Chang-Hong 4 is a medium altitude, long-range drone. It is capable of flying for up to 30 hours, at a range of 3,500 km, and at an altitude of 8,000 m, and carries a payload of 60 kg.[34] Developed by the China Aerospace Science and Technology Corporation, it is an updated variation of the earlier Chang-Hong 3, and is a rough equivalent to the US-made Reaper drone. Chinese scientists have suggested the Chang-Hong 4 will be a multi-role aircraft, with the optional capacity for targeted strike, but is more likely to be deployed for surveillance in harsher environments.[35] It is capable of carrying 2 AR-1 air-to-ground missiles, a laser ground missile comparable to the Hellfire munitions used by the Predator, and has also been depicted at air shows with the FT-5 precision guided "small diameter bomb", which is outfitted with a semi-active laser seeker for terminal guidance.[36] In August 2014, this was demonstrated in a live fire exercise, with the footage broadcast by Chinese state media.[37] Later interviews with Chinese scientists included claims that the Chang-Hong

4 could launch missiles from 5,000 m in altitude to hit ground targets within 1.5 m of accuracy.[38]

The Chang-Hong 4 is something of a rival to the Wing Loong within China's drone industry. There is contradictory information regarding the likely role of each drone in the PLA, and some defence sources have mistakenly grouped them together as the same vehicle.[39] Both vehicles have been described as the spearhead of China's unmanned exports; so far, the Wing Loong has been the technology of choice for foreign buyers, although Algeria has experimented with the Chang-Hong 4.[40] Given that the Wing Loong is an older and cheaper vehicle, it is more likely that it will be the mainstay of China's peace-time roles (in reconnaissance or "targeted killing" strike), while the Chang-Hong 4 is reserved for combat operations.

The BZK-005 (alternately "Giant Eagle" or "Sea Eagle") is a high/medium altitude, long range drone. It is capable of flying for roughly 40 hours at a maximum altitude of 8,000 m, a maximum range of 2,400 km, at a cruising speed between 150-180 km/hr.[41] The BZK-005 is designed for maritime surveillance and is currently used by the People's Liberation Army Navy (PLAN), capable of carrying a 150 kg payload, most of which is likely to be taken up by electro-optical, infrared, synthetic aperture radar and signals intelligence sensors. Jointly developed by the Beijing University of Aeronautics and Astronautics and Hongdu Aviation Industry Group, work first began on the drone in 2005, with images surfacing in 2009 of the vehicle sitting on a tarmac.[42] In 2011, there was a reported crash of the BZK-005 drone, discovered by local villagers who uploaded an image onto the internet before the crash site was cordoned off by the security forces.[43] Experts believe that the failure was due to a problem with the vehicle's guidance systems, and that at least some of the aircraft have since been grounded.[44] Despite this, a Chinese drone detected near Japanese air space in September 2013 was likely a BZK-005, suggesting that earlier problems had been resolved and that the aircraft was operational.

The Xianglong ("Soaring Dragon") is a long-range, high altitude reconnaissance drone. It is capable of flying for 10 hours, with a range of 7,000 km, a maximum flying altitude of 57,000 ft, and a maximum speed of 750 km/hr.[45] The Xianglong was first unveiled as a model in 2006, and images of the vehicle on the tarmac of Chengdu Aircraft Corporation's test ramp surfaced in 2011. After this, the aircraft design underwent significant changes to its length and wingspan, suggesting that initial tests had revealed problems with its aerodynamic profile.[46] Chengdu is likely partnering in production of the

Xianglong with Guizhou Aircraft Industry Corporation. Resembling the US-made Global Hawk surveillance drone, the Xianglong is believed to be designed for aerial reconnaissance, with additional sensors to designate maritime vessels for targeting by China's missile forces.[47] According to the US Defence Department, the Xianglong is also capable of carrying weapons.[48]

The WJ-600 ("Sky Hawk") is a jet-powered attack drone. Designed by the China Aerospace Science and Technology Corporation, it can reportedly fly at a maximum speed of 720 km/hr, with a sustained loitering speed closer to 108 km/hr, at an altitude of 10,000 m, at a 2,100 km range, but only for three and a half hours.[49] It is meant to provide a strike capability against hardened targets on the ground and water, includes two places to attach munitions, and could plausibly deploy the KD-2 missile, TB1 anti-tank missile, ZD-1 bomb, and FT-6 small diameter bombs. Launched via rockets, the drone is believed to use parachutes for landing.[50] The WJ-600 was first displayed at the 2010 Zhuhai Air Show in a video which depicted the projectile attacking a hypothetical aircraft carrier.[51]

The Lijian ("Sharp Sword") is an armed combat drone in the early stages of development, which resembles the US Navy's X-47B prototype. There are few technical specifications available on the vehicle, although there are rumours that it was reverse-engineered from Russia's Mikoyan Skat combat drone.[52] It has been jointly developed by the Hongdu Aviation Industry Group and Shenyang Aircraft Corporation, and was intended to replace the current batch of slow-moving, low-flying unmanned aerial vehicles in the Chinese military inventory. The Lijian was first reported in May 2013, and its first test flight was covered by the Chinese state media in November 2013.[53] This is China's first tail-less drone, and the available images of its airframe structure imply that it carries munitions in internal bomb bays. As with the X-47B, the Lijian is possibly a demonstrator vehicle for a future drone project. However, the Chinese state media has quoted experts who raised the possibility that it could be incorporated into aircraft carrier operations.[54]

The Anjian ("Dark Sword") is a conceptual model which has no equivalent in any publicly available drone project around the world. It is a supersonic vehicle which is capable of air-to-air combat, as well as ground strikes, and is believed to be the product of Shenyang Aircraft Corporation.[55] Some experts have noted that the characteristics of the Anjian are not optimised for signature suppression, suggesting that air combat is the focus, not suppression of air defences.[56] The design and placement of air intake vents suggest the Anjian is even being designed for "super-cruise" flying – or supersonic flight without the use of engine afterburners.[57] This likely makes it more ambitious than the latest combat models being developed by other countries. While the vehicle's technical specifications are not available, one authoritative study argues that the concept "should be treated with some scepticism given the secrecy that has surrounded the programme since its initial showing in 2006... it is given some credibility by the leading role Shenyang has played in China's stealth R&D".[58]

Beyond several other names which are infrequently mentioned as experimental vehicles, there are two more projects which merit attention. The Zhan Ying ("Warrior Eagle") is a conceptual model that is said to be designed for the suppression of air defence systems. It is depicted with a wing structure and materials that are capable of lower speeds to land on an aircraft carrier.[59] The "Sky Saber" is a combat drone, which was mentioned by the US Defence Department in its most recent assessment of China's military modernisation.[60] Beyond even these limited details, other drone concepts or ideas can be found on open-source forums which amount to little more than rumours.[61]

While many details are lacking, this list of projects points to an ambitious programme for indigenous drone production. Most of these prototype vehicles mirror equivalent US designs, suggesting that China has been playing technological "catch up", but not all: the Anjian and the Zhan Ying in particular appear to be truly novel concepts, which outstrip the known plans of competitor nations. While there is little indication of material progress, they are a reminder of China's ambition to pioneer new drone technologies in the future.

Strategic Environment and Training

For all this progress, China's drone capabilities are almost certainly handicapped by the inexperience and lack of training opportunities afforded to its military personnel. Despite the limited insight into China's drone industry, as well as the political and defence authorities overseeing development, there is almost nothing to indicate whether these vehicles are capable of undertaking military operations. As one recent study warns, "[w]hile China is seeking to advance stealth, electronic warfare, and lethal strike capabilities, demonstrated functions have been quite limited".[62]

The PLA is clearly interested in exploiting new possibilities opened up by drone research, but replicating technology will only provide so much benefit for a professional military. There remains the task of building expertise through intensive pilot training, familiarising controllers with new operating technology, educating analysts for processing data, and integrating the vehicles with battlefield information systems. Short of trial by conflict, the preparation and training in these areas are difficult to assess, but it is unlikely that China can achieve much in the short period of time since it began developing new indigenous technology.

Of the several issues which are conducive to research and development, China is deprived of a favourable strategic context in which to harness this prototype technology. The US military's global base of networks, support capabilities, and battlefield demands in Afghanistan and Iraq, permitted the quick adoption and experimentation of drone technology in the last decade.[63] By contrast, the PLA is more strategically confined in its potential for deployment, and lacks any recent history of adaptation to urgent contingencies. Surveillance drones have been reported as participating in maritime exercises, and there will be more roles for their use as China naturally expands its regional presence. But there does not appear to be any immediate operational demand for China's drones.

This is reflected in the limited deployment of China's drones thus far. According to one study, "The PLA's focus appears to be on employing UAVs for ISR [Intelligence, Surveillance, Reconnaissance] and for communications relay, in which forward-deployed UAVs pass information to command and control units (land-, sea-, or air-based)".[64] This is heavily skewed towards the more basic task of monitoring and protecting the country's interior and border areas, in line with Chinese strategic posture. In this arrangement, current drone models largely complement, rather than substitute, other military assets. Over time, this will make it more difficult to justify their cost, and at the very least, casts some doubt over the capacity for military innovation in the future. Even as the state of unmanned technology improves, the PLA may struggle to harness it for military advantage.

Deception and Information Warfare

Considering the example of China's military modernisation so far, another reason to be sceptical of its alleged progress in drone operations is the country's history of deception. Experts have long pointed to the influence of Sun Tzu over Chinese strategic thinkers, especially the ancient writer's preoccupation with the psychological component of rivalry.[65] Interpreted in contemporary terms, Sun Tzu's writings emphasise the use of calculated deception and manipulation of information to manoeuvre an enemy into unfavourable strategic territory. This is not so much a blueprint for war-fighting as it is a strategic tool for positioning; instead of prosecuting a battle, the challenge laid down by Sun Tzu, and embraced by Chinese military thinkers, is to induce the surrender of a competitor before war has begun. [66]

There is a compelling argument to be made that China's military has applied this lesson to its capability development, and that several high-profile aerospace projects have been publicised in an attempt to bolster its conventional deterrence. In recent years, a number of the People's Liberation Army's scientific achievements have been initially revealed on open source forums, usually from amateur observers on the internet who disseminate unconfirmed footage captured from near a local military base. While not betraying many useful details about new platforms or assets, this feeds into the narrative of a more militarily powerful China, leading to discussions on internet chat rooms and professional forums, only to prove premature after the typical developmental problems associated with rapid force modernisation set in.

This could be seen in the calculated dissemination of information surrounding the Jin-20 combat aircraft, a fifth-generation fighter.[67] In 2010, news of the Jin-20 leaked out in a series of photos and videos, most of which appeared to be imagery captured by local observers, who could have been prevented from loitering nearby if secrecy was paramount. Shortly after a visit by then US Defence Secretary Robert Gates in January 2011, the Jin-20 undertook its first, brief test flight, which raised questions about the pointed timing of this exercise, especially considering that some senior Politburo leaders (including now President Xi Jinping) were rumoured to have visited the test facility. One news source also quoted unnamed Chinese sources who suggested that the timing of the test flight may have been in retaliation for the news announced by the US military that it would be modernising Taiwan's fleet of F-16 aircraft.

Since that time, however, the Jin-20 has encountered the typical problems and delays associated with an experimental prototype. There are rumours that the aircraft was based on an abortive Russian model, and that the weak aerodynamic design inherited from this project was impeding the manoeuvrability of the plane.[68] The 2014 annual report by the US Department of Defence on the Chinese military acknowledged that the aircraft "faces numerous challenges to achieving full operational capability, including developing high-performance jet engines".[69] Indeed, several variations of the original model have been tested since 2010, with each successive vehicle boasting incremental improvements to the structural design and stealth capability.[70]

China's recent progress in developing drones is also likely to be as deliberately crafted as the Jin-20. The same process of excited gossip and fragments of information occurred during the test flight of the Lijian, after the secretive drone's maiden flight was captured in photographs that were released anonymously.[71] This would also explain the consistent drum of official news coverage, including the prominent role which has been accorded to drone models at anniversaries, exhibitions, and military parades in China.[72] Indeed, given that the PLA still resists international calls for transparency, the visibility of drone technology, and the expert testimony about their military role, may be a telling sign that they are not fully prepared for operational use. If not necessarily always untrue, the image being cultivated is a clear attempt to bolster China's conventional deterrence.

This is especially so considering the many challenges which were encountered in the US experience with unmanned technology. Not all the information about China's drone fleet is obviously the product of manipulation by the higher authorities; but at the very least, uncritical perception has glossed over many of the developmental constraints which inevitably lie ahead for the PLA. For instance, it is reported that the design and manufacture of a functioning propulsion system which could be used in unmanned aerial vehicles is still a technical hurdle for China's aviation industry. As with the Jin-20 fighter, current drone models rely on a Russia-made turbofan which lacks the radar-evading qualities needed to operate stealth aircraft.[73] For these reasons, the Anjian and Zhan Yiing unmanned concepts, which outstrip even the US military's plans, are still likely to be more hypothetical than substantive.

China's military and political leadership evidently believes that unmanned air power is valuable enough to justify the cost of pursuing these ambitious concepts. Even while experts are right to caution against the publicity hype and talk down the immediate threat posed by China's drones, this does not mean that the whole effort can be explained away as some kind of ruse. More likely, it suggests that senior officials in the PLA think there is prestige to be won in advertising these weapons, albeit prematurely, precisely because their future value is self-evident. While offering grounds for scepticism about China's military capabilities, this could also reflect the ambitious scope of China's military plans over the long run.

Most importantly, this indicates that China may have yet to make a serious choice regarding the most ambitious drone capabilities in the future. The image being cultivated by China's military and research industry is one of multi-directional development, with progress recorded across most conceivable vehicles and operational concepts, ranging from low-level tactical support to long-range strategic strike. Earlier tactical drones were abandoned, and some models under development appear to be more advanced than others, but this is apparently more a function of technical feasibility than any clear procurement decision made on high.[74] There appears to have been little effort to rationalise the drone projects under development, as there has been in the US. Indeed, while those organisations and aviation firms which specialise in strategic-level drones, like the Shenyang Aircraft Corporation, are "relative newcomers" in the field and expected to increase their sales, the established market for smaller vehicles is also expected to remain popular.[75]

The story is different for the PLA. While there are many drone projects receiving development funds, only a few of the vehicles have so far been incorporated for operational use. This is ultimately the best litmus test with which to assess the combat potential of new unmanned technology. To be sure, so long as it has replicated the American example, China's military and political authorities are likely to welcome investment and development. But given the ambitious objectives of the Anjian and Zhan Ying models in particular, choices may soon be made about the extent to which drones are used in the future.

China largely meets the criteria for fielding an advanced drone fleet, and is pursuing several models which closely mimic the American example of unmanned air power, as well as at least two models which differ from the US military's plans. The country's economic, military, and digital infrastructure is well suited to the rapid acquisition and deployment of unmanned technology, although some questions remain about its institutional capacity for experimenting and innovating with new drone models in the future. The next chapter examines the major trends in unmanned technology and science, to assess the challenges which lie ahead for China's future drone capabilities.

Notes

1. Defence Science Board, *The Role of Autonomy in DoD Systems*, US Government, July 2012 Report, p.71.

2. See, for example, Tai Ming Cheung, *Fortifying China: The Struggle to Build a Modern Defense Economy*, 2008.

3. Roger Cliff et.al, *Ready for Takeoff: China's Advancing Aerospace Industry*, RAND, 2011 Research Monograph, p.39.

4. Bradley Perrett, "Delays on Comac C919 Push First Flight to 2015", *Aviation Daily*, May 24, 2013, accessed at http://aviationweek.com/awin/delays-comac-c919-push-first-flight-2015

5. Cliff, et.al., n.3, p.27.

6. Beyond AVIC affiliates and subsidiaries, other corporations are listed in Kimberly Hsu et. al., *China's Military Unmanned Aerial Vehicle Industry*, US-China Economic and Security Review Commission, June 2013 Research Report, pp.9-13.

7. Ian Easton and Russell Hsiao, *The Chinese People's Liberation Army's Unmanned Aerial Vehicle Project: Organizational Capacities and Operational Capabilities*, Project 2049, March 2013 Research Report, p.7.

8. Tate Krasner, *Assessing China's Unmanned Aerial Vehicles: Policy, Development, Implementation, Capabilities, and Exports*, RSIS Research Manuscript, p.15, accessed at http://careeredge.bc.edu/wp-content/uploads/sites/4/2014/08/Chinese-UAV-Paper-Rewrite.docx

9. The tertiary institutes commonly mentioned are the Beijing University of Aeronautics and Astronautics, Nanjing University of Aeronautics and Astronautics, and Northwestern Polytechnical University in Shaanxi.

10. Daniel Houpt, "Civilian UAV Production as a Window to the PLA's Unmanned Fleet", Jamestown Foundation, *China Brief*, vol.12, no.4, February 2012 Report, accessed at http://www.jamestown.org/single/?tx_ttnews%5Btt_news%5D=39030#.VIWF1DGUeVM

11. Hsu et. al., n.6, pp.7-8.

12. See Houpt, n.10.

13. Edward Wong, "Hacking US Secrets, China Pushes for Drones", *New York Times*, September 20, 2013.

14. Easton and Hsiao, n.7, p.6.

15. Ibid., p.3.

16. Krasner, n.8, p.22.

17. Easton and Hsiao, n.7, p.11.

18. Aditi Malhotra and Rammohan Viswesh, "Taking to the Skies: China and India's Quest for UAVs", *Journal of Indian Ocean*, 2014, p.7.

19. Anand V., *China's Space Capabilities*, Centre for Air Power Studies, August 2012 *New Delhi* Research Paper, p.14.

20. Shashank Joshi and Aaron Stein, "Emerging Drone Nations," *Survival*, vol.55, no.5, 2013, p.59.

21. J. Michael Cole, "Beidou Satellites Raise Fears of Threat to Taiwan", *Taipei Times*, December 29, 2011.

22. Andrew Tate, "China Launches Latest of Military, 'Experimental' Satellites", *IHS Jane's Defence Weekly*, September 28, 2014.

23. Kranser, n.8, p.16.

24. See, for example, Richard Fisher, "Maritime Employment of PLA Unmanned Aerial Vehicles", in Andrew Erickson and Lyle Goldstein, *Chinese Aerospace Power: Evolving Maritime Roles*, 2012.

25. Scott Wilson, "Israel Set to End China Arms Deal under US Pressure", *Washington Post*, June 27, 2005.

26. Fisher, n.24.

27. Easton and Hsiao, n.7, p.11.

28. Christopher Bodeen, "China's Drone Program Appears To Be Moving Into Overdrive", *Associated Press*, May 3, 2013.

29. Hsu et. al., n.6, p.9.

30. Ibid., p.14.

31. See "Warplanes: Chinese UAVs of Arabia", strategypage.com, May 17, 2014, accessed at http://www.strategypage.com/htmw/htairfo/20140517.aspx

32. Wendell Minnick, "China's Unmanned Aircraft Evolve from Figment to Reality", *Defense News*, November 26, 2012.

33. See Houpt, n.10.

34. Minnick, n.32.

35. Xu Tianran, "Orders Taken for Chinese Drone", *Global Times*, November 15, 2012.

36. Minnick, n.32.

37. Gordon Arthur, "China Confirms CH-4 UCAV in PLA Service at 'Peace Mission 2014' Drill", *IHS Jane's Defence Weekly*, September 1, 2014.

38. "US Predator Not as Good as China's CH-4 Drone: Chinese Media", *WantChina Times*, September 3, 2014.

39. See "Unmanned Aerial Vehicles", Sinodefence.com, accessed at http://sinodefence.com/unmanned-aerial-vehicles/#CH4

40. "Algeria Evaluating Chinese CH-4 UAV", Defenceweb.com, accessed at http://www.defenceweb.co.za/index.php?option=com_content&view=article&id=33927:algeria-evaluating-chinese-ch-4-uav&catid=35:Aerospace&Itemid=107

41. Michael Sheridan, "China Builds Fast to Trigger Drone Race with America", *The Sunday Times*, September 15, 2013.

42. See "BZK-005", Airforceworld.com, accessed at http://airforceworld.com/pla/english/BZK-005-High-Altitude-Reconnaissance-UAV-China.html

43. Ibid.

44. Jerry Bonkowski, "China's Drones: Threat or Money Maker", *Asia Pacific Defence Forum*, June 28, 2013, accessed at http://apdforum.com/en_GB/article/rmiap/articles/online/features/2013/06/28/china-adds-drones

45. "Xianglong UAV Being Developed by China", Satnewsdaily.com, October 23, 2013, accessed at http://www.satnews.com/story.php?number=301313633

46. See "Soaring Dragon", Chinese-Military-Aviation.com, accessed at http://chinese-military-aviation.blogspot.com.au/p/uavucav-ii.html

47. See "Xianglong", Airforceworld.com, accessed at http://www.airforceworld.com/pla/english/Xianglong-UAV-china.html

48. Department of Defence, *Military and Security Developments Involving the People's Republic of China: 2014*, US Government, May 2014 Report, p.33.

49. See "WJ-600", Deagle.com, accessed at http://www.deagel.com/Unmanned-Combat-Air-Vehicles/WJ-600_a002594001.aspx

50. See "CASIC WJ-600 UCAV", Militaryfactory.com, accessed at http://www.militaryfactory.com/aircraft/detail.asp?aircraft_id=1031

51. Jeremy Page, "China's New Drones Raise Eyebrows", *Wall Street Journal*, November 18, 2010.

52. David Axe, "China's First Stealthy Killer Drone Takes Flight", Medium.com, November 21, 2013, accessed at https://medium.com/war-is-boring/chinas-first-stealthy-killer-drone-takes-flight-1766036badc0

53. "Successful First Test Flight for China's New Unmanned Stealth Drone, the Sharp Sword", *Associated Press*, November 22, 2013.

54. Zhao Lei, "China Conducts Test Flight of Stealth Unmanned Drone", *China Daily USA*, November 22, 2013.

55. Defence Science Board, n.1, p.70.

56. See "Anjian", Unmanned.com, accessed at http://www.unmanned.co.uk/autonomous-unmanned-vehicles/uav-data-specifications-fact-sheets/anjian-dark-sword-unmanned-aerial-vehicle-uav-specifications/

57. Fisher, n.24.

58. Easton and Hsiao, n.7, p.8.

59. Fisher, n.24.

60. Department of Defence, n.48, p.33.

61. Hsu et. al., n.6, p.5.

62. Krasner, n.7, p.28.

63. Lynn Davis et. al., *Armed and Dangerous? UAVs and US Security*, RAND, May 2014 Research Paper.

64. Hsu et. al., n.11, p.4.

65. See, for example, Fumio Ota, "Sun Tzu in Contemporary Chinese Strategy", *Joint Forces Quarterly*, vol.73, April 2014, pp.76-80.

66. Thomas Mahnken, *Secrecy and Stratagem: Understanding Chinese Strategic Culture*, Lowy Institute for International Policy, February 2011 Research Paper, p.21.

67. See, for example, "China's New Project 718/J-20 Fighter: Development Outlook and Strategic Implications", China Sign Post, January 17, 2011 Briefing Paper, accessed at http://www.chinasignpost.com/wp-content/uploads/2011/01/China-SignPost_18_J20-analysis_17-January-2011.pdf

68. "China's J-20 a Derivative of Russia's Withdrawn 1.44 Fighter", *Want China Times*, May 4, 2014.

69. Department of Defence, n.48, p.67.

70. Bill Sweetman, "J-20 Stealth Fighter Design Balances Speed and Agility", *Aviation Week and Space Technology*, November 10, 2014.

71. Imagery of the Lijian drone can be accessed at http://alert5.com/2013/11/21/photo-hongdu-lijian-stealth-uav-maiden-flight/

72. "UAV in Active Service Reveals PLA's Growing Interest in Military Robot", *Xinhua*, October 1, 2009.

73. Malhotra and Viswesh, n.18, p.6.

74. Trefor Moss, "Here Come...China's Drones", *The Diplomat* online, March 2, 2013, accessed at http://thediplomat.com/2013/03/here-comes-chinas-drones/?allpages=yes

75. Hsu et. al., n.6, p.14.

3. Future Trends in Science and Technology

The previous chapter demonstrated that China has a growing range of drone models and concepts under development. However, unmanned technology remains a scientific field undergoing rapid change, and developments in the future could quickly upend today's ideas about drone technology. This chapter surveys the general state of unmanned air power development, as well as China's agenda for drone research in the years ahead.

Several areas are likely to influence the future of the Chinese unmanned fleet, including autonomous processing, electronic warfare, expendability and affordability. All of these remain in a flux, and may produce technology surprises; indeed, it is impossible to predict how they could play out. But already it is clear that the PLA's most ambitious concepts regarding drone warfare are technically demanding, and almost certainly unrealisable in the near future. Instead, China is more likely to see progress in select air power roles for limited combat support. There is moderate scope for greater autonomy in military vehicles and over time, this will be reinforced by technology advances in commercial development, internal security, and defence exports.

Autonomy

Drones have received considerable public attention for their novelty, and few ideas are more dramatic than robots replacing human pilots in combat. China has announced at least two drone combat-capable models, which implies a greater reliance on autonomous thinking. This is a growing area of research for military technology more generally.

But autonomy does not necessarily translate into the total absence of human control over aircraft. Defined here as a capacity for "self-governing" or automatic behaviour, autonomous processing has already been harnessed for modern aircraft and sensors, but on a limited scale, with human pilots generally overseeing functions like diagnostics checks or targeting for weapons systems. These are repetitive tasks which take place within a well-modelled environment, akin to the motor skills of robotic technology in factories.[1]

The US military is particularly interested in this field. In a 2010 review of the future of technological challenges, the air force stated, "Increased use of autonomy – not only in the number of systems and processes to which autonomous control and reasoning can be applied but especially in the degree of autonomy that is reflected in these – can provide the Air Force with potentially enormous increases in its capabilities, and if implemented correctly, can do so in ways that enable manpower efficiencies and cost reductions".[2] But while more attention and resources are being directed towards the concept of autonomous technology, this is not leading to independently-thinking machines. Instead, it is relieving the cognitive burden on human operators by automating more tasks.

Indeed, a number of technically feasible innovations could easily free up more manpower in the years ahead. Take, for instance, the small-scale drones used for tactical reconnaissance by the US Army. These vehicles are managed by two personnel, who divide the responsibility for flying the vehicle and monitoring its video feed. With improvements to a drone's software functions, however, it would be possible to task a single drone for a specific mission profile, such as looking for an insurgent travelling through a mountainous region where the army personnel were based. After delivering these instructions to an automated computer receiving, the human controllers can let a computer weigh up the vehicle's flight performance characteristics, sensor capabilities, and optimised terrain conditions to plot a flight trajectory. Once underway, the same computer might use processing software to screen the raw footage being collected, and alert the human controller to anomalies or identified targets that were picked up in the mission.[3]

This can extend to the production of entirely new vehicle concepts. For example, a more autonomous drone may be dedicated to a surveillance and reconnaissance role in aerial combat. The ability to prioritise the collected data for transmission to human analysts would help relieve the bandwidth strain on digital networks, which has become a larger burden for drone operations. This would also reduce strain on the analytical workforce, which is struggling to keep up with the growing volume of sensor data that is monitored and processed.[4]

It is widely agreed that the science of autonomy is likely to deliver gains in efficiency across a range of war-fighting functions. As one report argues, greater autonomy "…will be enabled by continued progress in efficiency and miniaturisation of computer processing and power sources, and advances in machine learning and processing power".[5] This will likely result in a growing scope for unmanned initiative, whereby an autonomous drone takes more decisions on tasks where human judgement is not essential. Indeed, for the latest US drones, this kind of delegated authority already extends to take-off and landing, waypoint navigation, path planning, and rerouting upon loss of communications. As the Defence Science Board concludes, "The processing requirements for such autonomy software is well within the capabilities of today's laptops and embedded processors and is, therefore, ready for insertion into these systems".[6]

There are also signs that changes of this sort are underway. The US military is particularly interested in this field, as part of the recent "Better Buying Power" procurement reforms. These are meant to cultivate technological breakthroughs which can compensate for the diminishing lead in qualitative military power, and they identify autonomy as a priority investment.[7] By favouring an open-architecture structure for procured technology in the future, the scientific military community hopes to encourage third-party and private research innovation of the software. With a common operator system as the basis for experimentation, this would allow for regular software enhancements over time without the costly process of overhauling the basic model. This is fertile ground for quick, evolutionary progress in autonomous applications.

Artificial Intelligence

By contrast with graduated autonomy, the notion of artificial intelligence is an order of magnitude more complex. Instead of automated technology in a factory assembly line, this might be considered akin to the cognitive process undertaken by mobile robots used in planetary exploration, which are expected to navigate situations of far greater uncertainty.[8] In effect, this is autonomy extended to the maximum extent possible, where drones can make sense of novel problems and respond to surprises on the battlefield without the need for human guidance. If achieved, it would be truly revolutionary for air power doctrine, and pave the way for armed combat drones which could rival manned aircraft.[9]

This is almost certainly the most demanding aspect of unmanned technology in the future. It implies that control by a pilot, while optional, is unnecessary. If need be, an autonomous drone should be advanced enough to determine how its pre-programmed instructions can be applied in any kind of battlefield scenario, without necessarily thinking for itself to the point where it decides to abandon the mission. At present, the field of cognitive science is simply not capable of achieving this sophistication. The US Defence Science Board has identified several areas where the relevant technology and knowledge was lacking, including perceptual processing, planning and learning, human-robot interaction, natural language understanding, and multi-agent coordination.[10] This is also the most unpredictable factor for defence planning, as there is a considerable uncertainty among experts about whether advances in these fields can ever reach the point of combat superiority against human opponents. Whether the research can achieve this within any realistic timeframe, or at all, is impossible to say with any informed judgement.

But there are some enduring military considerations, which will weigh on this technology. At first glance, there are advantages to self-regulated artificial intelligence, and these could very well drive more innovation, slowly bridging the gap between theory and reality. Were human decision-making totally removed from drone thinking, for instance, the vehicle would benefit from an improved reaction time, be capable of experiencing gravitational force that would cause a human pilot to lose consciousness, and be less exposed to the risk of communications interference.[11] Indeed, some experts have claimed that extending the scope of autonomy to its maximum extent is necessary to withstand cyber-attacks.[12] According to this view, electronic warfare capabilities in the future could simply exploit the communications links of armed drones to hijack control of its computer; the only solution is to remove human oversight entirely, and trust in pre-programmed software.

However, this introduces problems for the concept of artificially intelligent drones. A fully remote drone severed from all communications links is effectively beyond salvage. Without even the most indirect degree of supervision by human controllers, minor glitches in software could lead to unnecessary casualty rates.[13] Already, there is some evidence of vulnerability in the US military's drone fleet: in March 2011, a Predator which had its ignition manually turned off, suddenly started its engine, in what technicians believed was related to data corruption.[14] A malfunctioning drone beyond control can only be disabled or shot down, which would be a particularly risky task if the vehicle is armed and programmed to defend itself. This poses a challenge for self-guiding artificial intelligence; whether it is resolved will depend on the volatile state of cyber capabilities and communications security in the years ahead.[15]

There are other weighty objections to artificial intelligence. A self-guiding autonomous drone without any communications link would be incapable of participating in most conceivable mission roles, because interaction with other military assets would introduce the risk of spoofing these vehicles. No matter how technically capable, self-guiding drones could be fooled by an enemy mimicking friendly behaviour, or generating false commands. These subversive attempts may be defeated, but as long as the legitimate controllers lack the means to oversee the drone's behaviour, it poses a fundamental problem for operations, whether in a combat or support role.[16] These security concerns are likely to buttress legal and moral objections about the possibility of "de-humanised" warfare.

Human-Robot Interaction

Given the formidable challenges ahead for the science of artificial intelligence, these problems cast doubt over the likelihood of self-regulating drones providing a reliable combat capability. Instead, a growing role for autonomous thinking alongside human cognition is a more likely alternative, with the prospect of manned-unmanned teaming appealing to experts, especially in support roles like electronic warfare.[17] This would involve a single human pilot controlling multiple accompanying drones in support of the mission. The chief advantage is that the requirements for processing, while still significant, are less onerous and raise fewer ethical and security issues.

According to the US Defence Science Board, there are two ways to understand the artificial intelligence demands for manned-unmanned teaming. These are described as "remote presence" and "taskable agency", and roughly correspond to the degrees of control exercised by the human pilot.[18] Remote presence involves a drone functioning as part of a pilot-centred system of vehicles, adding extra capability to a network which collectively assesses the environment, reducing the amount of information which needs to be comprehended by the human mind. By contrast, the taskable agency would involve a looser coupling of man and machine, with a pilot delegating specialised operational functions in the mission, leaving the drone the initiative to achieve its set objective. This might resemble something like the contemporary role of an air traffic controller: while interacting with autonomous pilots on nearby planes, the human cedes direct control in exchange for the ability to simultaneously track the progress of multiple vehicles and coordinate the broader movement of airborne activity.[19]

This could be the best potential for drones in the future, if only because it overlaps with other areas of defence research, in particular the concept of swarming. According to experts, swarming relies on multiple cheap vehicles functioning as an interdependent network.[20] This provides more operational redundancy, extends the area of coverage, and depending on the sensors and numbers, reduces the cost of aircraft losses. The shorter life-cycle implied in this technology would also permit more frequent innovation to the aircraft, and almost certainly provide more familiarity for the human pilots in training. Chinese presentations at the Zhuhai Air Show have already depicted several combat drones operating in tandem, suggesting that unmanned collaboration was on the country's research agenda.[21]

With more interest and investment directed into swarming, particular sub-fields are likely to develop more quickly than others. There is reportedly considerable progress being made in the process of automated cueing for situational awareness; the research into human interaction with computer vision processing has been described by the US Defence Science Board as "promising".[22] With greater improvements in this field, the perception of more events and objects can be delegated by human operators, reducing the likelihood of a pilot missing something in a cluttered environment.[23] Similarly, the sense-and-avoid capabilities for unmanned technology comprise an area that is also experiencing progress, and will be reinforced by commercial practices across a range of commercial industries, particularly in maritime navigation. Indeed, according to the US Defence Science Board, "The primary gap appears to be less in the fundamental but rather in hardening these solutions and integrating them with existing technologies and within socio-organisational constraints".[24] This suggests that improvements can be expected to occur in the near term, perhaps quite rapidly, with flow-on effects for the benefit of unmanned technology.

There is also evident interest among many countries, which are looking to incorporate teaming drones into mixed fleets with manned aircraft. France and Britain are funding feasibility studies for a European combat drone, but with the aim of only moderate improvements in unmanned technology, which is likely to fulfil an electronic warfare support role.[25] The PLAAF has reportedly begun exploring the science of interaction between unmanned and manned vehicles, and control links for human pilots.[26] Moreover, the science of manned-unmanned teaming extends beyond just aircraft operations, encompassing other types of units. For instance, the US Army has developed an air brigade for this very activity, which mixes Kiowa scout helicopters with Shadow-200 drones. Indeed, a 2011 experimental exercise matched several different manned and unmanned aerial vehicles, with the aim of integrating different sensors and information streams from multiple platforms.[27]

Electronic Warfare

To be sure, the concept of manned-unmanned teaming is not without complications. Another factor to consider is the threat posed by electronic warfare, particularly efforts at jamming or spoofing drone communications, which is certain to advance over time. Exactly how this balance plays out in the future is unknowable, but it will be a permanent risk for drone technology

It is reasonable to assume that all drone-human collaboration will rely on a communications link. This is to allow pilots to deliver instructions, as well as provide regular data for positioning, situational awareness, and internal system checks for unmanned vehicles. Indeed, the US Defence Science Board argues that, "…[r]egardless of the content of explicit communication between unmanned systems or a centralised server, robust network communication is essential for strongly and most weakly coordinated systems".[28] But this makes any manned-unmanned team inherently vulnerable to electronic interference. Some figures in China's drone industry have described this as a critical uncertainty for armed drones in the future, with one company executive conceding that "…[n]o one can guarantee the absolute safety of the data link connecting the drone and the control station, especially when the two sides at war have similar capabilities in electronic warfare and countermeasures".[29]

To be sure, the state of electronic warfare capabilities is an unpredictable factor for air power more generally. Heavily encrypted communications are difficult to break; they might be jammed, but other fail-safes built into the design of vehicles could reduce, if not wholly offset, this kind of threat. For instance, multiple data links could be attached to a drone for the sake of redundancy, so that even a targeted electronic attack against one device would not necessarily knock out the entire support system which enables unmanned flight. In addition to this, a heavier drone equipped with counter-measures might be capable of surviving at least some anti-air defences: indeed, one US military spokesmen has argued that the suite of electronic counter-measures available on an F-16 fighter would probably be sufficient for "even our current generation of RPAs [remotely-piloted aircraft]…they're surprisingly survivable in denied territory".[30]

The state of autonomy science will also play an unpredictable role. For instance, the US Defence Science Board describes "unaware" collaboration between drones which might plausibly withstand jamming efforts.[31] In this kind of network, which is analogised to an ant colony, drones do not fully comprehend the presence and activity of other drones. Instead, vehicles adhere to instructions which are patterned for an aggregate effect through limited monitoring of their surroundings. Researchers have experimented with this type of robotics for foraging, and are now considering its application for self-correcting minefields, which could detect any gaps or malfunctioning units next to them and adjust accordingly.[32] For air power, "unaware" processing in drones may provide a low-cost means of collaborative missions, with collision-avoidance software helping to keep drone swarms within a tight formation. With a limited or no signal needed, the vulnerability to electronic interception may also be reduced.

But this would also come at the expense of operational effectiveness. With only a small capacity for autonomous thinking on each individual drone, patterned behaviour is generally reliable but not especially flexible. This will stretch the capacity of a swarm to achieve complex objectives. The wider the operating environment, the more likely that surprises will occur, and more widely dispersed swarm activity will struggle to produce a cascading effect. While some progress in autonomy could offer relief against electronic warfare, it will likely be limited to simple tasks until more capable software processing is harnessed.

Affordability

Over the long term, it is impossible to predict the interplay between these many different factors. While the future will see greater autonomous thinking for drones, there are serious challenges ahead. The armed combat concepts for unmanned aircraft being developed by China, and some ideas put forward by drone advocates in the US, envision aircraft which conduct ambitious operations that are denied to manned alternatives, but the artificial intelligence requirements for these scenarios are difficult to satisfy without also risking greater vulnerability. As a result, over time, more viable models will probably scale down human pilot input into secondary functions, like monitoring aircraft sub-systems, while accompanying drones in a manned-unmanned team fulfil a greater support role. Even for these limited vehicles, the danger of electronic warfare will have to be overcome.

All of this calls into question a traditional advantage of drones, which is their expendability. In theory, a fleet of unmanned vehicles should provide more combat resilience, because aircraft without human pilots can maintain a higher rotation at lower cost, and can be replaced quickly if destroyed. This relies on the drone technology remaining cheap enough to risk some losses, but not the destruction of every vehicle. Against the threat of modernising anti-air defences and electronic warfare capabilities, the cheapest vehicles may well be too vulnerable, and their use could entail such high unit losses that the logic of expendability no longer applies.[33] At the very least, therefore, the computer processing demands for future drones will be much higher, so that they can avoid an unsustainable casualty rate. This is likely to require more sophisticated and potentially heavier equipment, and a correspondingly higher price for drone units, making operators more risk averse in combat.[34]

The fate of the US Navy's UCLASS drone is already a sign of cost trends in unmanned technology. It is now widely accepted among scientists and researchers that progress in drones will involve trade-offs in the design of the aircraft.[35] This, in turn, requires a decision about which specific capability to prioritise, and selecting a military advantage while also tolerating some weaknesses and vulnerabilities. This explains why the US military has at least deferred other combat drone projects for the moment, although the air force's plans for a future unmanned bomber were recently modified to make a pilot optional.[36] Even as the science of autonomy progresses, the prospect of fully-autonomous combat drones in the US military looks problematic without a vastly larger investment of resources.

State of China's Technology

How China acts on these issues has yet to be revealed. The PLA may see in the US example a warning of technological overreach, or it may believe its failure to press ahead opens up the way for seizing a competitive edge. But, at the very least, it is clear there are already shortcomings in the country's access to the unmanned technology in use today, and Chinese industry remains some time away from the wholly indigenous development and production of world-class drone technology. This is because in China, the aviation and aerospace sciences lag behind the American sector more generally, and because several key complaints repeatedly surface in interviews with Chinese experts and in the coverage of the state media.[37] Taken together, this gives an indication of problem areas which need to be addressed in the coming years before some of its more ambitious concepts can be indigenously built.

The principal hurdle for China's unmanned air power is widely believed to be the development of engines. Some propulsion systems designed for use in China's latest aircraft are now being produced by the country's manufacturing sector, but they reportedly have severe problems with reliability.[38] Across China's military and civilian aerospace sector, most turbofan and turboprop engines are usually imported from Russian suppliers.[39] While some knowledge will be gained by local operators and developers through these imports, there are significant differences between the design specifications for high performance military aircraft and commercial vehicles. This is also likely to be the case for most operational drones in the future, which will require advanced engines to support a larger weight, including high-powered sensors and munitions, while also being capable of improved manoeuvrability.

Similarly, China has yet to engineer a drone model with a structural design that is optimised for stealth. The 2014 annual report on the Chinese military by the US Department of Defence highlighted this area as a development priority, saying that the PLAAF "views stealth technology as integral to unmanned aircraft, specifically those with an air-to-ground role, as this technology will improve the system's ability to penetrate heavily protected targets".[40] Thus far, the most advanced unmanned aerial vehicles like the Lijian ("Sharp Sword") which have been publicly revealed, benefit from a partial improvement in radar-avoiding technology, with the lack of a visible tail on the end that would reflect radar waves. But amateur imagery also shows that the drone's modern jet engines are still exposed, which would likely leave it vulnerable to detection to modern air defence systems.[41] This is a shortcoming which extends to the rest of China's manned aircraft fleet. For the moment, unlike the US Navy's UCLASS project, the Chinese aviation industry appears to be content with building new vehicles on incremental progress in stealth capabilities.

There is also a likely shortfall in sensor and communications technology. After all, this equipment must be of a particular kind which can be attached to light-weight drone vehicles, and is, therefore, not widely available and difficult to access. For instance, it was reported in early 2009 that China had attempted to purchase video and thermal imaging cameras from a South African defence manufacturer of unmanned surveillance vehicles, but that its efforts ended up deadlocked in negotiations.[42] Along with other reports of select purchases from abroad, this was almost certainly a sign of the Chinese industry trying to obtain technology for drones.[43] Much of this appears to have been reverse-engineered from foreign models, which are purchased or captured in a more primitive form and updated later. For example, several days after the Iranian military claimed to have grounded a US surveillance vehicle, some experts noted the presence of a visiting team of Chinese scientists and technicians, who reportedly inspected the vehicle and were suspected of taking some key components back to Beijing for analysis.[44]

The flight software for onboard computers is also likely to be much less advanced than the US equivalent, given the state of the technology in China's aviation industry. For instance, the 2013 annual report on the Chinese military by the US Department of Defence notes that "solid state electronics and micro processors" as well as "guidance and control systems" are key foreign technologies which are targeted by Chinese operatives to compensate for indigenous shortcomings.[45] Few public details are available to independently assess this claim, but the remote operating terminals for the Wing Loong drones exhibited at the Zhuhai Air Show only possessed three monitor screens instead of the standard five used by the US Air Force for Predators and Reapers[46]; this suggests a poorer capacity for navigation and guidance control.

Structural Advantages

However, there are also several advantages for China's drone industry which the People's Liberation Army can exploit to gradually redress its shortcomings over time. As one study cautions, "[a]cquisitons that leverage strong Chinese balance sheets and weak Western markets can change the dynamics in a disruptive fashion even in an industry with high entry barriers like aerospace".[47] With these underlying processes fuelling the increasing use of drones, it may be that at some point in the future, China enjoys globally competitive unmanned capabilities.

Many of the associated technologies for military drones – such as autopilot software, satellite navigation, inertial guidance, digital mapping, and collision avoidance – can be as readily used in civilian aviation.[48] As a result, the non-military part of China's drone industry will likely drive a greater share of innovation over time, because drones are well positioned for rapid expansion in China's commercial markets. The country's aviation industry can be expected to develop in line with the rising level of per capita GDP, as the option of faster transport appeals to wealthier Chinese individuals and businesses, particularly for cargo freight where time costs (for items like fresh produce) outweigh ticketing costs. But there remain some hurdles for the industry, particularly in airfield infrastructure: whereas the US had more than 18,000 general purpose aviation airports in 2008, the equivalent number in China was only 71. Another shortage is in police and maritime search helicopters: all of the rotary wing projects in China collaborate with foreign firms, or are concerned with older Western models, and the total production capacity of the industry is limited.[49]

For these and other niche roles, cheaper commercial drones already appear to be filling a commercial gap. For example, the Aircraft Owners and Pilots

Association estimates that there are now roughly 10,000 professional drone operators in China, and around 300 separate enterprises involved in aviation work, including a growing number of unregulated start-up firms.[50] Not all of these are registered and approved for flying; indeed, one media account suggests that up to 80 percent of current drone activity is not yet authorised.[51] As civil aviation regulations adjust, this private sector enthusiasm for drones is a valuable resource for government agencies like the State Oceanic Administration, which already saves costs on maritime patrols by sub-contracting drones instead of naval vessels.[52] With dense urban centres strung along China's coastline favouring shorter transport flights on average, a greater share of civil unmanned aviation is likely to be taken up by drone services over time.

China's well-documented environmental problems are giving rise to further uses for drones. Unmanned technology is well suited to civilian response efforts because it limits human exposure in emergencies and natural disasters, alongwith providing a more persistent capability. For instance, civilian drones have already been sent in response to the 2013 earthquake in Sichuan and the 2014 earthquake in Yunnan, helping to map the disaster zones for Chinese rescue teams on the ground.[53] This is likely to be only the first of many civil surveillance functions performed by drones, from nuclear radiation detection, fire and flood monitoring, to more routine functions like resource mapping.[54] All of these will be a boon to local research and development, and may even come to eclipse the scale of the PLA's drone operations. Indeed, in August 2014, the Chinese Academy of Surveying and Mapping operated a civilian drone in the air for 30 hours, setting a new record for the country's use of drones well beyond what the military has been known to achieve.[55]

China is also poised to invest heavily in drone technology for law enforcement and public security. Already, civil unrest in China's rural provinces, especially some local elements of extremism associated with the Muslim Uighur community in Xinjiang, has resulted in the use of drones. This is an issue which has received greater prominence, with terrorist attacks, broader civil unrest, and police crackdowns taking place more recently.[56] After an outbreak of violence in July 2014, an order for surveillance drones by local security officials was fast-tracked by China's Aerospace and Technology Corporation, and company technicians were quickly dispatched to assist with the handling of the new technology.[57] This example highlights the central role which law enforcement also plays in the military-commercial nexus for research and development. Indeed, the police agencies and security organisations which operate drones in

China are served by the Nanjing Research Institute on Simulation Technique, otherwise known as the 60th Institute in the People's Liberation Army's GSD.[58]

More so than in most other countries, China's aviation industry will benefit from a legal framework that is more favourable to intrusive aviation. This is because there is a range of political challenges for the central government in Beijing, and virtually no constraints on domestic surveillance for law enforcement and national security. In addition to combating violence, civil authorities have used drones to monitor the enforcement of pollution restrictions on industries and local governments, which often try to boost output by circumventing environmental regulations, as well as the activity of corrupt officials and criminal networks involved in the drug trade.[59] While Chinese air space is mostly controlled by the military authority, it is widely anticipated that regulations for low-altitude air space will be loosened by the government to allow for the growing civil aviation market.[60] This will particularly benefit low-altitude, short and medium-range drones.

In a sign of this trend, the Chinese State Ocean Administration has announced plans to build a network of 11 drone bases along the country's coastline to assist in environmental monitoring and maritime surveillance. The original plan was to develop these bases by 2015, which would encompass airborne patrols near the disputed Senkaku/Diaoyu Islands; since then, few details have been revealed, although one recent study has identified at least two bases currently under construction, in Dalian and Yingkou.[61] Already, in late 2011, provincial government officials in Dalian confirmed the use of surveillance drones to monitor the water and islands near the Korean Peninsula, in what could be an attempt to crack down on illegal Chinese fisheries in South Korean territory, or preparation for an unexpected surge in North Korean refugees over the maritime border.[62] And individual drone flights have also received publicity, including several in Zhejiang province during 2014 which reportedly patrolled a geographic area which was 75 sq km in size.[63]

Beyond this, it is likely that in the years ahead, broader progress in China's military aviation sector will spill over into unmanned research and development. For instance, the most advanced drone models being pursued by China (the Anjian and Zhan Ying), are being developed by the Shenyang Aircraft Company, and are said to be drawing on that company's existing resources in stealth technology and fighter capabilities.[64] This is particularly true of the most ambitious drone concepts, such as long-range bombing, which will rely on advanced technology that can be employed on manned aircraft.[65] But even in the near term, one Chinese media report flagged improvements in flight control, data processing, navigation, power supply, and launch-recovery systems.[66]

Diplomacy

There is a strong diplomatic component to China's drone industry. As with other weapons systems, a growing sales profile will help reduce the cost of drone research and development over time. At the 2012 Zhuhai Air Show, for instance, Chinese developers said they were "interested in exporting" drones, and that this was "why we're displaying them over here".[67] This is a sensitive part of Beijing's foreign policy, with little verification, but several military partners are suspected of operating Chinese-made drones, or slight variations of the original vehicles. One study has also suggested that the China National Aerospace Technology Import and Export Corporation (CATIC) operates as the business development arm for the country's drone exports, although this remains unproven.[68]

Several countries are believed to have purchased Chinese drones. After an unidentified drone was spotted in Uzbekistan's air space in 2012, suspicion quickly fell on China; experts noted that the vehicle's dimensions appeared to match the specifications of a Wing Loong.[69] There are also rumoured deals with the United Arab Emirates and Uzbekistan, which have not been dampened by the refusal of both countries to comment on the speculation.[70]

China's military ally Pakistan possesses an unknown number of drones, which it reportedly uses for surveillance in counter-terrorism and law enforcement missions. First unveiled in November 2013, the Burraq and Shahpar vehicles were said by Pakistani sources to have been built indigenously, but their striking resemblance to the Chinese made Chang-Hong 3 raised suspicion that the aircraft, or at least their major components, were developed in collaboration with China's aviation sector.[71] While Pakistani officials maintain these drones are unarmed, the original Chang-Hong 3 is capable of carrying weapons. While popular resistance within Pakistan to the US-led drone campaign in the tribal regions has led the government to push back against American policy, this new batch of drones likely offers greater scope for independent military action in the future.

Saudi Arabia is a confirmed recipient of China's Wing-Loong drone.[72] However, the nature of this revelation is telling: Chinese media reported on the issue only after the news was published by a Russian military website, which referenced the Saudi prince who reportedly managed the drone deal.[73] This agreement remains shrouded in uncertainty, with no details offered about the number of vehicles, or the negotiated price. Saudi Arabia is said to have turned to the Chinese after its request for a US-made Predator was first rejected by its American partner.[74]

China has also supplied drones to the Algerian military for testing, as well as offering training to the controllers. Algeria is interested in acquiring vehicles for tracking and intercepting the Maghreb-based extremist groups emanating from conflict zones in Mali and Libya. Media reports suggest that military units practised on the drones for several months, and that while at least two vehicles crashed during landing at air force bases, local officials were still keen to purchase the technology.[75]

Thus far, China's known penetration of the global drone market is limited to these countries. While the US and Israel possess more sophisticated unmanned technology, their commercial prospects are restrained by concerns about the risk of proliferation in unstable regions, particularly the Middle East and Africa. Indeed, the technical limitations of Chinese drones probably make them more financially appealing for the poorer countries in these regions, which are sure to be tempted by cheaper products without political strings attached. The Wing-Loong drone, for instance, is reportedly marketed at roughly USD 1 million, which is much cheaper than the estimated price of a Predator.[76] This competitiveness also extends to associated munitions, with China equipping its drone vehicles with the HJ-10 Red Arrow missile, which costs roughly USD 70,000 compared to more than USD 100,000 for the Predator's Hellfire missile.[77]

Drone exports may also provide a new avenue for China's defence diplomacy. According to one Chinese media account, during an August 2014 military drill between the members of the Shanghai Cooperation Organisation (SCO), the PLA practised live-fire exercises with unmanned aerial vehicles attacking ground targets, although the specific vehicle was not mentioned in reports.[78] Military officials claimed that the unidentified vehicle had not missed a shot during its practice exercises.[79] There is also the potential for greater collaboration with Israel's defence industry, as the two countries continue to explore the joint development of business jets. Given Israel's advanced unmanned technology, and its past dealings with the PLA, one study remarks that "if it [Israel Aerospace Industries] transfers UAV-related technologies, it will not surprise anyone".[80]

Since the leadership ascension of Xi Jinping, China's plans for multilateral engagement have perceptibly accelerated, with a network of governance organisations being proposed by Beijing as an alternative to the traditional American-led order. This includes the newly proposed Asian Infrastructure Investment Bank, a Regional Comprehensive Partnership for trade, and the revamped Conference on Integration and Confidence Building Measures in Asia, which played host to Xi's widely-quoted speech on China's conception of security in early 2014.[81] At present, these multilateral arrangements represent diplomatic posturing more than they do substantive cooperation between members, but this may change in time. In any case, the shallow nature of these forums is no impediment to further drone sales; if anything, the desire to broaden the scope of military collaboration may encourage Beijing to pursue drone exports as a means of securing greater prestige and, ultimately, strategic influence.

The Near Future

Over the long term, all these factors will reinforce China's aggressive development strategy for unmanned technology. But, in the meantime, for all the interest generated by futuristic models, China faces serious obstacles in developing technology for air combat. Trends in scientific research point to a future where autonomous drones are more likely to fulfil a supportive role, alongside manned aircraft, not one of replacing them.

This is not to declare artificial intelligence for air power impossible, or to suggest that China will not pursue this option in any case. As with the historical example of the Cold War space race, it may be the case that the prestige and imagined spin-off technology associated with this idea motivates China, along with the US, to devote more resources in the future. But at present, the indications are that this will be a greatly increased sacrifice, and that it is likely to encounter risks from cyber and electronic warfare which calls into question its feasibility. Indeed, there are several others areas of drone technology which China first has to master before indigenously producing combat drones.

At the very least, this is a reason for defence planners to focus on the technology under development, with a critical eye towards anything too ambitious. Several underlying factors place the country's industry at an advantage over the long term, but this is also best suited to less sophisticated equipment and less capable vehicles for export. The next chapter examines the diplomatic and strategic consequences for regional security which are likely to arise from these developments.

Notes

1. Defence Science Board, *The Role of Autonomy in DoD Systems*, US Government, July 2012 Report, p.31.

2. US Air Force, *Report on Technology Horizons: A Vision for Air Force Science and Technology, During 2010-2030*, US Government, July 2010 Report, p.105.

3. This is detailed in Defence Science Board, n.1, pp.31-32.

4. One estimate by the Defence Science Board is that 20,000 human analysts will be needed to maintain the status quo. Ibid., pp.83-84.

5. Samuel Brannen, *Sustaining the US Lead in Unmanned Systems*, Centre for Strategic and International Studies, February 2014 Research Report, p.5.

6. Defence Science Board, n.1, p.32.

7. See Undersecretary of Defence Frank Kendall's speech to Air Force Association 2014 Conference, September 17, 2014, accessed at http://www.defense.gov/Speeches/Speech. aspx?SpeechID=1884

8. Defence Science Board, n.1, p.31.

9. Michael Byrnes, "Nightfall: Machine Autonomy in Air-to-Air Combat", *Air and Space Power Journal*, May-June 2004, pp.48-75.

10. Defence Science Board, n.1, p.54.

11. Ibid., p.33.

12. Joshua Foust, "Why America Wants Drones That Can Kill Without Humans", *Defense One*, October 8, 2013.

13. Mark Gubrud, "No, Killer Robots aren't Secure", Medium.com, October 16, 2013, accessed at https://medium.com/i-m-h-o/no-killer-robots-arent-secure-a7c6981915e1

14. Craig Whitlock, "Remote US Base at Core of Secret Operations", *Washington Post*, October 26, 2012.

15. Gubrud, n.13.

16. James Jinnette, "Unmanned Limits: Robotic Systems Can't Replace a Pilot's Gut Instinct", *Armed Forces Journal*, vol.147, no.4, November 2009, pp.30-32.

17. Brannen, n.5, p.10.

18. Defence Science Board, n.1, pp.45-47.

19. Ibid.

20. See, for example, Paul Scharre, *Robotics on the Battlefield, Part II: The Coming Swarm*, Centre for New American Security, October 2014 Research Report, pp.5-7.

21. Richard Fisher, "Maritime Employment of PLA Unmanned Aerial Vehicles," in Andrew Erickson and Lyle Goldstein, *Chinese Aerospace Power: Evolving Maritime Roles*, 2012.

22. Defence Science Board, n.1, p.35.

23. Byrnes, n.9, pp.56-57.

24. Defence Science Board, n.1, p.36.

25. Cyril Altmeyer and Tim Hepher, "France, UK Award Contracts to Study New Combat Drone", *Reuters*, November 5, 2014.

26. Ian Easton and Russell Hsiao, *The Chinese People's Liberation Army's Unmanned Aerial Vehicle Project: Organizational Capacities and Operational Capabilities*, Project 2049, March 2013 Research Report, p.12.

27. Paul McLeary, "US Army Presses Ahead on Manned-Unmanned Teaming", *Defense News*, April 30, 2013.

28. Defence Science Board, n.1, p.53-54.

29. See, for example, Li Yidong, Vice General Designer for the Chengdu Aircraft Design & Research Institute, cited in Xu Tianran, "Orders Taken for a Chinese Drone," *Global Times*, November 26, 2012, n.97.

30. Sydney Freedberg, "Drones Need Secure Datalinks To Survive Vs. Iran, China", *Breaking Defense*, August 10, 2012.

31. Defence Science Board, n.1, p.51.

32. Ashlee Vance, "The Self-Healing, Self-Hopping Landmine", *The Register*, April 11, 2003.

33. This is a problem for air power more generally. See Thomas Hamilton, "Expendable Missiles vs. Reusable Platform Costs and Historical Data", RAND, October 2012 Research Paper.

34. Prem Mahadevan, "The Military Utility of Drones," CSS Zurich Analysis, no. 78, July 2010, p.2.

35. Stew Magnuson, "Navy Program at Center of Drone Survivability Debate", *National Defense Magazine*, September 2014.

36. Kris Osborn, "Air Force Plans Next Gen Drone Fleet", Defensetech.com, June 19, 2014, accessed at http://defensetech.org/2014/06/19/air-force-plans-next-generation-drone-fleet/

37. See, for instance, Wang Yangzhu from Beijing University of Aeronautics and Astronautics cited in Zhao Lei, "Nations Look to Buy Drones from China", *China Daily USA*, June 20, 2013.

38. James Dunnigan, "Chinese Engines Travel A Trail of Tears", strategypage.com, October 7, 2014, accessed at http://www.strategypage.com/dls/articles/Chinese-Engines-Travel-A-Trail-Of-Tears-10-7-2014.asp

39. Roger Cliff, et. al., *Ready for Takeoff: China's Advancing Aerospace Industry*, RAND, 2011 Research Monograph, p.116.

40. Department of Defence, *Military and Security Developments Involving the People's Republic of China:2014*, US Government, May 2014 Report, p.66.

41. Stephen Trimble, "New Chinese Advances in Tailless UAV Designs Revealed", *Flight International* online, May 14, 2013, accessed at http://www.flightglobal.com/news/articles/picture-new-chinese-advances-in-tailless-uav-designs-385842/

42. Andrei Chang, "China Seeks South Africa's UAV Technology", *United Press International*, January 13, 2009.

43. Russel Hsiao, "Advances in China's UAV Program", Jamestown Foundation, *China Brief*, vol.10, no.19.

44. David Cenciotti, "Chinese Delegation Currently in Iran to Copy the US Stealthy RQ-170 Drone Captured in 2011", *The Aviationist* online, August 16, 2012, accessed at http://theaviationist.com/2012/08/16/chinese-delegation-rq170/

45. Department of Defence, n.40, p.49.

46. David Cenciotti, "Inside the Pterodactyl UAS: Sneak Preview of China's Predator Clone Mobile Ground Control Station", *The Aviationist* online, November 20, 2012, accessed at http://theaviationist.com/2012/11/20/pterodactyl/

47. A.K. Tiwary, "Unmanned Aerial Vehicles in China", *Indian Defence Review*, vol.28, no.1, January-March 2013, p.22.

48. Manjeet Singh Pardesi, "UAVs/UCAVs: Missions, Challenges, and Strategic Implications for Small and Medium Powers", RSIS Working Paper no.66, May 2004, p.18.

49. These details are cited in Cliff et. al., n.39, pp.9-30.

50. Lai Yuchen, "Blue-sky Thinking Colors China's Drone Industry", *Xinhua*, August 17, 2014.

51. Chun Han Wong and Kersten Zhang, "China Drone Operators Face Criminal Charges After Aggravating Air Force", *Wall Street Journal*, China Real Time blog, October 23, 2014, accessed at http://blogs.wsj.com/chinarealtime/2014/10/23/china-drone-operators-face-criminal-charges-after-aggravating-air-force/?mod=WSJBlog

52. Minnie Chan, "State Oceanic Administration Plans Drone Patrols for Coastal Areas", *South China Morning Post*, August 30, 2012.

53. "Drones Dispatched to Capture Images of Quake-Hit Regions", *Xinhua*, April 20, 2013.

54. Aditi Malhotra and Rammohan Viswesh, "Taking the Skies: China and India's Quest for UAVs", *Journal of Indian Ocean*, 2014, p.6.

55. James Dunnigan, "China Breaks a UAV Record", strategypage.com, August 20, 2012, accessed at http://www.strategypage.com/dls/articles/China-Breaks-A-UAV-Record-8-20-2014.asp

56. See, for example, Simon Denyer, "China's War on Terror Becomes All-Out Attack on Islam in Xinjiang", *Washington Post*, September 19, 2014.

57. Didi Kirsten Tatlow, "China Said to Deploy Drones After Unrest in Xinjiang", *New York Times* Sinosphere blog, August 19, 2014, accessed at http://sinosphere.blogs.nytimes.com/2014/08/19/china-said-to-deploy-drones-after-unrest-in-xinjiang/

58. Kimberly Hsu et. al., *China's Unmanned Aerial Vehicle Industry*, US-China Economic and Security Review Commission, June 2013 Research Report, p.12.

59. Jennifer Duggan, "China Deploys Drones to Spy on Polluting Industries", *The Guardian*, March 19, 2014.

60. Hsu et. al., n.58, p.14.

61. Tate Krasner, *Assessing China's Unmanned Aerial Vehicles: Policy Development, Implementation, Capabilities and Exports*, RSIS Research Manuscript, accessed at http:careeredge.bc.edu/wo-content/uploads/sites/4/2014/08/Chinese-UAV-Paper-Rewrite.dxx, pp.21-22.

62. "Chinese Province Starts Drone Patrol off Waters Near N.Korea", *Chosun online*, December 1, 2011, accessed at http://english.chosun.com/site/data/html_dir/2011/12/01/2011120101081.html

63. Wang Qian, "Drones Survey Islands in Zhejiang Province", *China Daily*, July 31, 2014.

64. Hsu et. al., n.58, pp.10-11.

65. Lynn Davis et. al., *Armed and Dangerous? UAVs and US Security*, RAND, May 2014 Research Paper, p.5.

66. "China's UAV Into a Period of Fast Progress", *People's Daily Online*, February 8, 2013, accessed at http://eng.chinamil.com.cn/news-channels/china-military-news/2013-02/08/content_5213125.htm

67. Quoted in Jeremy Page, "China's New Drones Raise Eyebrows", *Wall Street Journal*, November 18, 2010.

68. Easton and Hsiao, n.7, p.7.

69. Stephen Trimble, "Spotted: A Mystery (ok, maybe Chinese) UAV in Uzbekistan", flightglobal.com, February 20, 2012, accessed at http://www.flightglobal.com/blogs/the-dewline/2012/02/spotted-a-mystery-ok-maybe-chi/

70. Christopher Bodeen, "China's Drone Program Appears to be Moving into Overdrive," *Associated Press*, May 3, 2013.

71. Sara Sorcher, "Pakistan Wants Drones, and It Doesn't Need America's Permission to Get Them", *National Journal*, May 15, 2014.

72. Zachary Keck, "China to Sell Saudi Arabia Drones", *The Diplomat* online, May 8, 2014, accessed at http://thediplomat.com/2014/05/china-to-sell-saudi-arabia-drones/

73. "Saudi Arabia Signs Deal for China's Pterodactyl Drone", *Want China Times*, May 6, 2014.

74. Krasner, n.61, p.27.

75. "Algeria Evaluating Chinese CH-4 UAV", Naval Open Source Intelligence blog, March 12, 2014, accessed at http://nosint.blogspot.com.au/2014/03/algeria-evaluating-chinese-ch-4-uav.html

76. Estimates generally range from USD 5 to 20 million. See, for example, Joe Cochrane "At Asia Air Show, Plenty of Competition for Sales of Drones", *New York Times*, February 16, 2014.

77. Krasner, n.61, p.26.

78. Beth Stevenson, "Iranian and Chinese UAVs Used in Regional Military Drills", flightglobal.com, August 28, 2014, accessed at http://www.recreationalflying.com/threads/iranian-and-chinese-uavs-used-in-regional-military-drills.123325/

79. "China's Drone Blasts Off Missile in SCO Anti-terror Drill", *Xinhua*, August 26, 2014.

80. Tiwary, n. 47.

81. Nick Bisley, "The Real China Challenge: Beijing's Blueprint for Asia Revealed", *The National Interest* online, October 27, 2014, accessed at http://nationalinterest.org/feature/the-real-china-challenge-beijings-blueprint-asia-revealed-11550

4. Strategic Options and Future Risks

With China beginning to field more military drones, there is growing speculation about the way these will be used in the future. But while some information is publicly available on drone procurement and development processes, exactly how the PLA intends to harness this technology for its military doctrine and operational deployments is unknown.[1]

Unmanned technology is subject to unpredictable changes, and innovations are likely to continue, surprising defence planners in the years ahead. But there are at least several operational uses which are already the topic of speculation, and if successfully implemented, could have a substantial impact on the military balance and security dynamics in China's surrounding region.

Building on the technological problems and development opportunities detailed above, this chapter explores the possible roles which unmanned air power might play in China's defence posture. While by no means definitive, it offers a sketch of the likely military roles in which drones could be deployed by China, and illustrates some of the remaining hurdles to be overcome before more ambitious concepts can be achieved.

Strategic Effects

There are many types of unmanned technology beyond air power, and China can naturally be expected to diversify its drone capabilities over time. For instance, there is already considerable scope for the incorporation of armed drones into Chinese ground units. This is because the US military is trained in combined arms and manoeuvre warfare, and struggles to distinguish a friendly drone from an opponent's vehicle at short range; moreover, there has been little doctrinal preparation for a contingency involving enemy drones attacking land forces, with air defence systems straining to detect low-level aircraft in a cluttered surveillance picture.[2] As a result, the PLA may incorporate more tactical drones into its ground formations. While not necessarily as compelling, there are also promising uses for micro-drones, underwater and surface maritime drones, and rotary wing helicopter drones.

However, the capacity for these concepts to affect regional security in peace-time remains to be proven.[3] For unmanned ground technology, the inherently limited potential for long-range operation minimises the adjustment to the strategic posture which rival militaries need to undertake, beyond slight alterations to operational deployment in training exercises. Similarly, micro-systems are an area of great interest for military planners, but they are also greatly limited in range, and hardly qualify for the description of aircraft.[4] Along with rotary wing helicopter drones, and for the moment, maritime drones, it is hard to foresee these weapons having any serious impact on the regional balance of military power, short of conflict.

By contrast, even in times of peaceful diplomacy, the exercise and positioning of air power through surveillance and readiness has a more lasting effect on the politics of international security. The prospect of drones being added to this dynamic has given rise to some anxiety among proliferation experts: as one study argues, "…although these new weapons will not transform the international system as fundamentally as did the proliferation of nuclear weapons and ballistic missiles, they could still be used in ways that are highly destabilising and deadly".[5]

But it is important to place this concern about proliferation in the context of the Indo-Pacific region. China's drones will be operated within this security landscape, and this fact imposes several constraints, as well as presenting opportunities, which will shape the country's approach to unmanned technology. The following categories represent some of the more plausible options available to the People's Liberation Army in the near future.

Expanded ISR and Data Support

China looks set to exploit its growing fleet of drones for intelligence, surveillance, and reconnaissance functions, particularly in the waters of East and Southeast Asia. Not only are several vehicles, notably the BZK-005, well suited to this kind of role, but they would provide a significant tool for maritime operations. Following its war-fighting doctrine of "localised war under information heavy conditions", the PLA is developing long-range, precision-strike weaponry which can threaten the military assets of any other country in a growing radius from China's shoreline, as well as blunt an attempted intervention by the US military in a local conflict. The first condition for this kind of military campaign is a detailed surveillance picture of the surrounding region. Medium and long-range drones can provide another layer of intelligence collection assets, adding to the country's growing suite of over-the-horizon radar systems and satellites.

One study speculates that China's expanding fleet of drones "could be useful for detecting, locating, tracking, and targeting high-value fixed and mobile targets – such as US Navy ships – throughout the Western Pacific" as well as "perform[ing] ISR [intelligence, surveillance, reconnaissance] on fixed and mobile targets on Taiwan and in the Taiwan Strait" during war-time.[6] Already, the PLAN is sponsoring research which looks at unmanned helicopters as guidance platforms for long-range strikes.[7] Moreover, the need for situational awareness will only become more pressing as China continues developing a networked military force, with a connected system of ground platforms like mobile artillery, aircraft (including next-generation fighters and the upgraded bomber fleet), and warships all capable of cueing and launching guided munitions.

There are several operational advantages for the use of drones in this context. Without a pilot in the cockpit, even the most basic surveillance vehicles would enjoy greater endurance and loiter time. At the same cost as a manned equivalent, this would position a greater number of surveillance platforms on station, offering more redundancy against anti-air defences on ships and land. Alternatively, this could extend the reach of surveillance: most estimates of the operating range of the country's new drones are between 2,000 and 3,000 km from China's coastline, albeit with less loitering time.[8] Unlike ground-based radar installations and reconnaissance satellites, a fleet of drones is essentially mobile, in that it can be deployed in any direction at short notice, complicating a hypothetical enemy's targeting strategy. With a dedicated surveillance role, the drones would not require many other capabilities, and, thus, could be produced relatively cheaply, allowing for a large force which can absorb higher losses.

At first glance, the strategic consequences of this deployment do not appear to be overly concerning. The drones would have a serious impact in conflict, but even then they would likely be confined to a supportive role. Were surveillance drones to track foreign vessels and slow-moving aircraft throughout China's immediate region in anticipation of a sudden outbreak of conflict, however, it may have worrying implications for escalation stability in a diplomatic crisis. For instance, the US military would well be concerned at the visible exposure of its naval vessels and assets to Chinese targeting intelligence, secured by drones. Depending on the context, this could compress the time when a vessel being tracked could use its weapons before being attacked, perhaps encouraging counter-moves to evade detection, including jamming efforts against the drone as a last measure. A high concentration of surveillance drones shadowing vessels in peace-time might also produce unnecessary tension[9], although given the scarce deployment of surveillance drones so far, China's military decision-makers seem to understand how this could be construed as provocative behaviour.

Beyond this, low-observable drones could be utilised for covert reconnaissance of foreign territory. North Korea's example is a good indication of how a country might opt to utilise relatively unobtrusive aircraft to cross over land borders, without triggering the alarm of an opponent's air defence systems. But this relies on the use of unsophisticated tactical drones which can avoid detection, and it is unlikely that the PLA would gain much strategic advantage from these vehicles. Were more sophisticated sensors installed on these drones, the potential loss arising from unreliable pre-programmed navigation equipment is likely to be too high. Any serious improvement to the guidance controls to reduce this liability would entail changes to design that make the vehicles more detectable by air defence systems.

In addition to surveillance and reconnaissance, unmanned technology is likely to play a greater role in transmitting communications between various elements of the PLA. Indeed, the Chinese Navy is already experimenting with this option: prototype drones have been relaying data from deployed units to command and control centres; and since 2011, the South Sea Fleet has been training with unidentified fixed-wing drones for communications support.[10] This could also extend to so-called "pseudolites", or cheaper vehicles temporarily operating at extremely high altitude which can perform a communications function similar to dedicated satellites.[11] Given the emerging interest in space warfare with modern weapons threatening satellites, temporary alternatives in the form of drone pseudolites could be an attractive means of building redundancy into China's military communications networks. At least one known PLA study has explored this possibility, describing "robotic sub-satellites" which can loiter at up to 50 km from the earth's surface.[12]

This is an area of research wherein China is more likely to overtake the US military at some point in the distant future. American drones were not designed primarily for survivability, but instead to provide as extensive support as was possible to ground operations in the Middle East and South Asia. Unlike the US military, there is more insistence within the PLA on deploying future vehicles in a highly contested environment. As a result, reducing radar detection and cloaking communications are said to be receiving more emphasis in recent years.[13] These platforms are generally inoffensive, and do not appear to be problematic for regional stability.

Loitering Strikes

Equipped with armed drones, China has already considered emulating the US precedent of targeted strikes in foreign territory. Unlike many other countries, China has the infrastructure for cross-border operations, and faces several transnational risks to its national security, from splinter extremist groups to smuggling and criminal networks. Within the limits imposed by human analysts and the quality of its imagery sensors, the PLA is capable of undertaking a drone bombing campaign.

Drones have achieved their greatest impact in this kind of scenario.[14] With longer endurance for surveillance, a quieter profile which can achieve surprise in the evening dark, and the deniability afforded to unmanned technology, they can offer direct access to remote problems which lie beyond sovereign territorial borders. China is largely surrounded by politically weakened neighbours, like North Korea, Mongolia, Laos, Myanmar, Nepal, and a host of Central Asian states with little capacity to effectively and transparently manage their border areas. Moreover, few, if any, of these countries are likely to develop a robust air defence system; instead, security ties will more likely offer Chinese air power a permissive operating environment. Already, some reports have speculated on the likelihood of China deploying drones in Kazakhstan to track down extremist sanctuaries fomenting Uighur-led violence in Xinjiang province.[15]

Drones were reportedly floated as a possible option to assassinate Naw Kham, the leader of a criminal network in the Mekong delta who smuggled drugs across the Chinese border and raided maritime vessels. After consolidating his trade through hostage taking and piracy, Kham's associates murdered 13 rival traffickers who were also Chinese nationals in 2011. This provoked outrage in China, which began coordinating efforts with Thai and Burmese authorities to hunt Kham and suppress his criminal network. According to the head of the narcotics bureau in China's Ministry of Public Security, the possibility of a drone strike against Kham was considered at this stage, but dropped in favour of a manhunt on the ground. The abandoned plan would have called for a drone to drop 20 tonnes of dynamite on Kham's mountainous compound, where he was largely shielded from ground patrols by the local authorities.

The story attracted much attention when it first surfaced, but few facts have come to light since then. Crucially, the rationale behind the decision to abandon the plan has never been comprehensively explained. It could be that the political leadership in Beijing feared a backlash in international public opinion, that they judged the available technology to be unreliable, that they wanted Naw Kham arrested and tried for publicity value in China, or that they feared alienating the Burmese government.[16] For whatever reason, the option was said to be quashed by the higher authorities, suggesting that political, not operational issues, were at the forefront of thinking. This caution may not hold in every circumstance, but it has established a high bar for future drone activity by China.

The weight of international opinion is almost certainly more pressing than it was in 2011. What began as a fascination with drones in the early years of the Afghanistan War has increasingly turned to unease and criticism after the Obama Administration ramped up its covert bombing campaign. Public opinion in global forums is increasingly energised by the issue of drone warfare: there is a range of organisations, forums, and civil society campaigns agitating for the control and regulation of autonomous weapons. If not altering the substance of drone operations to satisfy its most trenchant critics, the US has modified its behaviour under the pressure of this scrutiny: it openly acknowledges a role in extra-territorial strikes, and has attempted to establish plausible legal grounds for this activity. Its reading of domestic and international law, which permits activity specifically targeting extremist groups associated with Al Qaeda, seems to have been motivated by warnings that other countries would soon begin the same practice.[17] This suggests that the Obama Administration is mindful of the need to appease public opinion, and that community pushback against breaches of sovereignty is now a factor to be reckoned with.

The policy of denial once offered by drone technology is no longer sustainable for a world power conscious of its image. This presents a complication for Chinese security policy which didn't exist several years ago. With more UN investigations underway, the international resistance to drone operations is perceptibly hardening, and China would face critical scrutiny if it undertook its own strikes. One study argues that "within the Chinese academic community, scholars have suggested UAV use for domestic surveillance, law enforcement, and noncombat tasks near China's borders, but few have considered its use overseas, fearing international criticism".[18] This is also a sensitive area for Chinese policy-makers, because the country has long taken a principled stand on the inviolability of sovereignty, criticising foreign intervention.[19]

At best, there would be a high risk of collateral damage and political backlash for bombing strikes in the future. If nonetheless embraced by China, this option would have to be narrowly limited to avoid the alienation of neighbouring countries. This level of detailed targeting was allegedly possible in the Naw Kham incident, but the same circumstances will be difficult for China to replicate. For instance, the Chinese Ministry of Public Security had formed a task force to track Kham, stationing more than 200 officers in Laos, Myanmar, and Thailand. In order to collect the evidence for prosecution, these agents conducted reconnaissance in the area and laid several traps to apprehend Kham.[20] These failed, but reportedly went some way towards developing a human intelligence network on the ground, which could provide surveillance of Naw Kham's zone of operations near the Chinese border, and quietly penetrate his network of sympathetic locals. China also appealed to the Burmese regime in Naypyidaw, which applied military pressure by assaulting the outposts in Kham's network, slowly encroaching into his territory, and reducing his freedom of movement.

Without the support of local collaborators, which offers a permissive aerial environment and a network of intelligence operatives to collect targeting information, a signature drone strike is not really feasible. Host governments without the capacity to patrol their border regions may openly invite China's intervention. But in the areas where this might take place, few smaller neighbours are exclusively dependent on Chinese power: Russia maintains strategic influence in Central Asia, and India and the US in Southeast Asia. Even if these local governments were willing to cede sovereign control of their air space, and China was prepared to take it, the strategic reality of multiple interested powers is likely to act as an added restraint.

It is possible that China nonetheless opts for this method in the future. Counter-terrorism has received more emphasis recently and President Xi Jinping places great emphasis on "New Silk Road" diplomacy which will increasingly bind Chinese economic and social stability to the internal affairs of neighbouring countries.[21] As a growing power with areas of civil unrest within or across its borders, China will be inclined to exercise more strategic weight in its region. But, at present, the balance of international factors seems likely to weigh against the use of drone strikes.

Diplomatic Pressure
Another option available to China would be the aggressive deployment of drones for diplomatic effect among regional competitors. Without the need to place a human pilot at risk, unmanned technology is an ideal tool for intimidation

and pressure. Whether armed with munitions or not, reconnaissance flights by drones can provide continuous presence in a region's disputed areas, even though the risk of a brief incursion into foreign territory is always present. These deployments could be used to test the resolve of a competitor to uphold its sovereignty, or slowly erode the security assurances from allied partners.

To some extent, this behaviour is already on display. Chinese maritime vessels have begun enforcing China's claim to shoals and reefs in the South China Sea, as well as the disputed Senkaku/Diaoyu Islands, following a strategy of low-level harassment against foreign vessels. Unmanned technology has not yet been used for this campaign, although a suspected Chinese BZK-005 drone was first detected near Senkaku/Diaoyu in late 2013, prompting the Japanese Self-Defence Force to scramble a group of F-15 fighter aircraft to patrol the area.[22]

If China wants to apply more diplomatic pressure in the future, these kinds of flights could be expanded and intensified. In the unlikely event that maritime relations soured beyond diplomatic negotiation, China might even use armed drones for isolated instances of violence to ward off third-party intervention. This could take the form of a single attack against civilian vessels of a target nation.[23] Additionally, the PLA could opt for provocative unmanned flights as a bargaining tactic in a diplomatic stand-off. This would be particularly useful in situations over land, where manned aircraft could not be risked without incurring serious political liability. This might also be the right gesture to signal displeasure at intransigent client-partner states like North Korea, which has chafed under Beijing's influence in recent years.[24]

This is not to suggest that these options would be sound policy; they run serious risks. But aside from maritime vessels, China generally lacks a coercive tool for regional bargaining which also offers sufficient "plausible deniability" to de-escalate any crisis. China is inhibited by its doctrinal reliance on strategic strike against military rivals, because the PLA lacks clearly delineated options for escalating coercion before the outbreak of conflict: ballistic missiles hold nearby ships and military bases at risk, but a single demonstration shot or a launch confined to a particular area will be hard for an opponent to discern from a broader attack.[25] As with nuclear weaponry, any effort to take one step brings with it the risk of unleashing a general war, particularly as military plans now envision conflict extending into space, with further risk to early warning systems.

Drones offer a better instrument with which to incrementally apply diplomatic pressure abroad. Whereas maritime vessels are limited to island disputes and risk head-on collision with rival vessels, a surprise flight of tactical drones could avoid physical confrontation, while still conveying a political message. If this gambit fails, the absence of a pilot leaves open the possibility of blaming mechanical faults or an unidentified third party; and this is arguably no longer a feasible option for China's maritime vessels in the South China Sea.[26] Drones are likely to be interpreted as less offensive to national prestige than army troops crossing into foreign territory, as has been the case with Indo-Chinese strategic relations. Moreover, because surveillance vehicles are relatively cheap to produce, they can be used on a larger scale for a demonstration effect, potentially in multiple locations at the same time. As with Russian military drills ramped up across Eurasia in the aftermath of the Ukraine crisis, China may step up provocative flights in the knowledge that its message will be understood by neighbouring countries. The added advantage is that all of these vehicles are, in the final analysis, expendable.

While not all equally probable, each of these scenarios does contain a risk of escalation to military confrontation. Drones can't receive, let alone respond to, visual or radio warnings from enemy aircraft, so the easiest and safest option for signalling to the opposing force is by disabling or destroying its vehicle. There is still a degree of political uncertainty about whether the use of violence against drones would cross a "red line" that demands a response. In this vein, the US military's experience of drone surveillance in the Persian Gulf is a concerning precedent: for many years, the Iranian Air Force has regularly observed US military aircraft undertaking regional patrols near its air space, and has not risked military activity beyond its sovereign territory. But in November 2012, Iranian aircraft fired on a US Predator drone, claiming the vehicle had breached Iran's sovereign air space. This was disputed by the US military, which maintained that it was flying in the regular patrol area, and prompted the Chairman of the Joint Chiefs of Staff to announce a "measured response" to guard American military assets in the future, including plans for manned aircraft to accompany drone surveillance flights for added protection.[27] Whereas drones were viewed as a safer alternative than manned aircraft, the Iranian example suggests their presence may perceptibly lower the threshold for violence, escalating the situation only if they are attacked.

In China's case, however, there is a clear interest in avoiding a precedent which rivals might also exploit in the future.[28] Maritime vessels are an attractive option for China because rival claimants to disputed islands in the South China Sea are generally unwilling to risk violent escalation, nor can they field as large a number of civilian trawlers to match China's asymmetric strategy. The same does not apply to drones: the frequent deployment of cheap unmanned aircraft could be an effective tactic for Vietnam, Japan, and the Philippines to frustrate China's hope of consolidating exclusive control over disputed territory. A similar deployment undertaken by larger military powers, such as the US or India, would play on the anxiety of a PLA which is already sensitive about the ongoing surveillance of its coastal installations. In recent years, a key source of US-China tension has been the US Navy's aerial and maritime presence near the Chinese submarine base on Hainan Island, which is the focal point for incidents between the two militaries.[29] If China hopes to establish a sanctuary free from external interference, it is not in its interest to establish a precedent of reckless drone surveillance.

There is some basis for a regional understanding on unmanned vehicle conduct which can avoid inadvertent escalation. Indeed, after several years of uncertainty, the Japanese government confirmed that it will shoot down drones which violate its sovereign territory. The absence of surveillance drones in China's campaign over the Senkaku/Diaoyu Islands might be taken as a sign that this threat has worked so far, with the PLA not willing to risk any provocative behaviour when red lines on military activity are clearly established. If the problem of drone intimidation emerges, other regional powers could easily adopt this position, reassuring China that they will not conduct drone flights. There is no guarantee that Chinese drones will not be used to violate territorial boundaries in the future, but a predictable and limited military response in these circumstances is much less likely to begin a spiral towards armed conflict.

Decoy Attacks

These concepts have been widely discussed, and much of the technology is ready for peace-time deployment. In a hypothetical conflict, however, there are several other options for the use of armed drones that China may consider over time. The strategic impact arising from these hypothetical weapons is harder to predict, and in some cases, is unlikely to ever materialise because the concept is not technologically feasible. But they are also the subject of great interest for military thinkers, and, thus, merit analysis.

One of the more likely developments for the PLA could be a larger number of decoy vehicles, which could be used in a massed campaign against heavily defended air space. Some Chinese military writing has reportedly explored the use of unmanned anti-radiation and combat vehicles, along with spoofing and electronic warfare methods, to disable and destroy enemy air and missile defences.[30] This kind of coordinated air campaign was pioneered by the Israeli military in the 1982 air battle in the Bekaa Valley in Lebanon, which commenced with drones emitting fake signals to distract Syria's radar defence systems, followed by manned aircraft to attack the newly revealed locations of the Syrian weapons.[31]

With advances in technology lowering the cost of unmanned flight, the idea of decoy vehicles which could generate these effects in a conflict zone is widely discussed. As one US Department of Defence Study suggests, "[j]ust flooding the air space with simple UAVs flying random patterns would create the equivalent of a flock of geese at the end of a commercial runway".[32] In China's case, this would place a premium on its cheaper models, which could be mass produced by a wide range of production facilities at short notice, helping to negate its traditional shortcomings in electronic and radar systems. While a growing number of J-6 fighter aircraft have been converted into unmanned platforms by the PLAAF for targeting practice, these might also be harnessed for decoy activity.[33] A hypothetical contingency might involve the use of drone air power to degrade Taiwan's sophisticated air defence system, before launching an amphibious assault on the island. One study which reviewed Chinese scientific literature regarding unmanned technology has suggested that planning should mostly focus on using swarms to penetrate the defensive shield surrounding US military aircraft carriers.[34]

Drones are evidently the only platform suited for this kind of initiative. There are several reasons why a fleet of decoy vehicles would become a more compelling option for China over time. An unmanned combat fleet would almost certainly cut down the cost of pilot training; and with a reduced dependency on human personnel, a larger force could be maintained at higher levels of readiness at the same cost.[35] This would favour the PLA in any long, drawn-out campaign, where attrition begins to impact on enemy personnel numbers, even as more drones could be manufactured at a faster replacement rate. Supported by other capabilities that inflicted losses, the large-scale deployment of decoy drones would complicate the war-fighting strategy of an opponent's air force. Against an enemy naval force, a large-scale deployment of decoy drones would also exhaust the magazine capacity of maritime vessels, imposing a proportionately higher financial cost on anti-air systems, which has long been a complication for the US Navy's efforts at missile defence.[36]

In order to achieve this, drones would have to be light and cheap, maintain robust sensors for targeting, a minimum weapons payload, and limited autonomous processing for airborne coordination. One study argues that "[s]warms of MAVs equipped with sensors and miniaturised warheads are theoretically capable of attacking high-value targets such as radars and launchers of SAM sites" because "micro-manufacturing and nano technology could provide an exponential leap in microminiaturisation for weapons, sensors, and platforms".[37] In all likelihood, given their relative expendability, it would be better for China to build on the self-destruct model of the Harpy drone, rather than equip each reusable vehicle with munitions.

But while feasible, this capability would be somewhat limited in application. Unmanned vehicles are generally more expendable than manned aircraft, but this does not make the concept affordable across all contingencies. For instance, the Chinese military doctrine does not assume a protracted conflict in the future, but instead aims for a short, localised military engagement.[38] A sustained conflict may play into the hands of potential rivals, such as India and the US, which are located far beyond Chinese waters in East Asia, allowing these countries to deploy their naval strength to blockade China's energy imports from the Persian Gulf. To be sure, Chinese ideas about hypothetical conflict may alter over time, or adjust to new socio-economic realities in the energy trade. There is also no guarantee that Chinese drone acquisition will flow coherently from the military doctrine. But at the least, it would seem that air power geared towards attrition is not an ideal option for the PLA.

There are other circumstances where a drone swarm tactic would be less effective than its proponents claim. For instance, a network of cheaper drones is likely to experience more resistance against a mobile layered air defence system, where more expendable anti-air platforms are placed at a greater range from the defending area, degrading a drone fleet over time as it proceeds towards the target. In order to unmask the location of concealed enemy weapons, any drone swarm must also be capable of mimicking the radar signature of more advanced vehicles, or jam defence systems after detecting them. This will make it more difficult to restrict the cost of these vehicles. Moreover, the already extensive missile range of the PLA makes it unlikely that high-value enemy vessels, such as aircraft carriers, would be stationed within easy range of a cheaper, less-capable drone fleet.

The dilemma for an unmanned swarm is that a longer range is almost certainly incompatible with expendability. Over greater range, swarm drones might be released by heavier manned platforms, including transport aircraft or air-capable warships, which could also offer some in-theatre guidance. But it could be difficult to house sufficient numbers of swarm vehicles on these platforms, while also making sure that the drones possess the onboard equipment to confuse enemy air defences, and withstand electronic interference from a rival force.

By contrast, a more effective deployment of decoy vehicles would probably be confined to military activity close to China and its drone control stations. According to one study, "An expendable UAV attacking a fixed objective does not need sophisticated target-finding technology or a long-range satellite link", and this is likely to be the case with well-surveyed military emplacements near the border.[39] Without the requirement to travel over long distances, the vehicles could arguably remain cheap to produce in large numbers, permitting the saturation of ground targets with many vehicles. While the technology has not yet proven capable, this is a hypothetical risk for China's neighbours, and arguably of particular concern to India, given the nature of the Himalayan terrain which separates the two countries. As one study argues, "It is not easy to conceal and camouflage effectively" ground-based military assets in the region, leaving them vulnerable to attack from any decoy or saturation attacks, especially if the vehicles are stealthier and quicker than manned aircraft.[40] Variations of the WJ-600 jet-powered drone may produce this capability for China in the future.

Unlike the development of surveillance drones which can shadow enemy naval vessels, decoy drones are unlikely to accentuate the risk of crisis instability, as China's large arsenal of ballistic and cruise missiles already poses a similar kind of threat. What it might accomplish is a retargeting of these weapons against other targets further from the conflict zone, with implications for missile defence efforts among target countries.

Combat Strike Roles

A more transformative option for China would be to continue developing its prototype combat drones, Anjian, "Sky Saber", and Zhan Ying. Rather than being designed for sacrifice in a decoy attack, these armed vehicles would contest for control of the battlefield, engaging and destroying military assets like rival aircraft, maritime vessels, and air defence systems. The concepts are necessarily more speculative, because the associated technology is still uncertain, but some basic problems and requirements can be teased out for analysis.

There is a number of potential attack roles which armed drones could offer the PLA. One widely discussed option is to directly target and destroy aircraft carriers.[41] The growing arsenal of cruise and ballistic missiles is already designed to perform this role, although at a longer distance from China's coastline; armed drones with greater endurance and loitering time might be able to extend this range of firepower even further. This imposes some limitations, because the need to maintain light weight means that drones can only be equipped with a limited number of munitions, and these may not be powerful enough to sink a large vessel. However, even a lightly armed vehicle could still threaten to disable air-wing operations on a carrier, and extend to nearby vessels: according to a 2012 Pentagon study, beyond the aircraft carriers themselves, an enemy drone attack "could be extended to rear echelon supply convoys and other combat support assets which have not had to deal with an airborne threat in generations".[42]

A potential tactic is for drones to directly contest manned aircraft for air control. At first glance, this seems like an even more ambitious concept, because of the demands placed on tactical fighters for speed and manoeuvrability. But it is unlikely that a combat drone would be capable of prosecuting an attack against an aircraft carrier without also being prepared to defend against the vessel's air wing and anti-air artillery. This type of drone would also presumably receive more interest over time as a defensive measure against bombers and fighters.

Unlike decoy drones, these vehicles would require greater weapon carrying capacity and low-observable flight technology, attributes which already match with the structural design sketched out for the Anjian. But in addition to this, these combat drones would need to be capable of vastly improved computer processing for navigating the battlefield. Importantly, these vehicles could probably not be remotely operated by human controllers on ground stations, because this would require a live video feed which is subject to electronic disruption. Without the need to send high-resolution imagery in real time, however, the electronic vulnerability of drones is considerably reduced. This implies that the drone must perform many more cognitive tasks, as the only humans who could continuously oversee its operations are accompanying pilots with a line of sight communications uplink.

This poses a fundamental technical challenge for unmanned tactical aircraft. The range of weapons and sensors on an aircraft carrier in particular makes it well suited to undertaking defensive electronic warfare measures to disrupt this collaborative teaming. It may be hypothetically possible for the PLA to field a combat drone which is sufficiently autonomous to withstand such electronic measures, while receiving only limited supervision of its decision-making, but this seems extremely unlikely within the foreseeable future. Whether over land or sea, the likely response from China's neighbours will involve the development of more accurate and powerful electronic warfare capabilities, increasing the technological hurdle for China's drones to coordinate effectively.

A less challenging alternative may be for the deployment of drones which are themselves capable of engaging in offensive electronic warfare. Instead of aiming to impose air control over manned aircraft, China's drone prototypes could provide combat support in a manned-unmanned team arrangement. Depending on the state of these attack capabilities, they might target fighter aircraft at a further distance, or hone in on support aircraft like refuelling tankers. The latter option would arguably be more in keeping with the PLAAF's traditional focus on attacking the enemy's infrastructure, like air bases and communications nodes.[43]

This is certainly an area of interest for China. According to one report, "PLA technical studies have repeatedly discussed operational concepts whereby electronic warfare UAVs are deployed in tandem with unmanned precision strike platforms, in some cases blurring the line between the two".[44] In fact, the China Electronic Technology Corporation has a research institute which is exploring swarming tactics and electronic warfare, including jamming.[45] This might take the form of additional jamming pods attached to larger drone vehicles, akin to the upgrade which produced the EA-18G "Growler" aircraft by combining electronic warfare capabilities to a combat fighter design.[46]

Anti-Submarine
A less ambitious operational role could involve these drones targeting submarine vessels. Drones equipped with sonar buoys would be able to assist in the hunt for submarines, by loitering over contested maritime zones and relaying their sensor data to calculate the underwater routes of enemy vessels.[47] Alternatively, the growing fields of moored sonobuoys being laid by the PLAN could eventually be structured as mobile sensors, channelling data to anti-submarine warfare capable drones.[48]

This is another area of interest for Chinese military thinkers. According to one report, the PLAN is "known to be exploring the possible applications of VTUAVs (Vertical Take-off UAVs), including their use in anti-submarine warfare".[49] Indeed, Chinese scientific journals have published articles on search algorithms which could be used by processors onboard drones to track submarine locations, by integrating data from multiple sources.[50]

Drones may prove more suited to this operational role than aerial combat. Loitering time and larger fleets of vehicles allow for a more comprehensive sweep of maritime waters, while the cognitive challenge is likely to be less imposing than duelling human pilots. Moreover, anti-submarine warfare remains a relative weakness compared with other regional navies.[51] As a result, anti-submarine drones would be filling a capability gap that doesn't really exist to the same extent for aircraft carriers or manned aircraft.

Strategic and Tactical Bombing

Another possibility for China is to build a dedicated bombing drone. While it is too early to say with any certainty, this may be the purpose of the Zhan Ying concept, or a spin-off model in the future. A cheap and more expendable fleet of drones could be stationed on China's aircraft carriers, threatening tactical strikes in amphibious operations. Alternatively, a larger and heavier drone, with internal bomb bays for stealth, could travel at high-altitude into a hostile environment and deliver a payload against political and strategic targets.

Carrier-based drones have been discussed widely, within the US military and in China.[52] At first glance, this seems like an appealing option for the PLAN, if only because unmanned technology could offset the limited hitting power of its new Liaoning carrier. Smaller drones would permit a greater number of vehicles to be stationed on this prototype carrier, extending the operating range and sidestepping the usual teething process of manned air wing operations on vessels at sea, which can take decades of painful experience to master.[53] This is an area of evident interest, with Chinese military and science journals exploring the technology of carrier-borne systems, and has the advantage of mimicking the US military's UCLASS project. Additionally, communications might be easier to maintain if the drones were to operate as part of a carrier battle group, with other aircraft in a data relay support role.

But this raises questions about storage and munitions payload. If the Liaoning or a future carrier were to be equipped with tactical bomber drones, these would not only have to be compact enough to be stored onboard in sufficient numbers, but also be capable of taking off from the vessel's flat top. Given the long timeline expected of this project, the underlying technology may shift drastically in the years ahead, making this capability more feasible. But as one study points out, any "[i]mprovements in electronics will not change the fundamental physics of delivering large quantities of munitions to long distances".[54] As a result, carrier-borne drones will find it difficult to carry the degree of firepower offered by rival systems. For instance, China's submarines and mobile land-based missiles are already capable of threatening Indian military installations, with high odds of estimated survivability.[55] This will inevitably extend across the Pacific Ocean in the years ahead, with the PLAN soon to possess submarine-launched ballistic missiles.[56] Unmanned bombing concepts would probably offer only marginal strategic utility on top of this.

The PLAAF is also said to be interested in developing unmanned technology for long-range bombing missions.[57] Tellingly, this kind of penetrating capability has also been flagged as a concern by the US military, with the Pentagon's Defence Science Board conceding that "[h]igh altitude systems, such as the Global Hawk UAV, could pose a problem as the United States has not encountered a manned threat in that regime since the Soviet-era MiG-25 Foxbat".[58] With their large geographic land mass and economic and military infrastructure, India and the United States might both be the preferred targets of this type of drone technology. If built to operate at higher speed and altitude, this vehicle might not need to be as capable of tactical manoeuvre against fighter aircraft, which could potentially reduce the need for the most sophisticated processing software for self-guidance.

However, it is not clear exactly why this type of vehicle would necessarily be unmanned. Removing the human pilot may gain some additional time for flight, but manned bombers like the US military's B-2 can already fly continuously for more than 30 hours. More so than any other combat vehicle, an autonomous bomber operating deep in enemy territory, presumably during a large scale conflict involving space assets, cannot be expected to rely on a communications link. This runs the risk of cyber subversion: strategically weighty decisions about targets in foreign air space would have to be ceded to pre-programmed software, which could always be tampered with before deployment. The danger of losing control over this powerful weapon casts doubt over its future development

Implications for Arms Control

In addition to military contingencies, China's future drone capabilities also raise questions for the global arms control framework. This is particularly so, as China does not belong to the relevant instrument for regulating the proliferation of drone technology, the Missile Technology Control Regime (MTCR).

Designed to halt the spread of delivery systems that could be mated with weapons of mass destruction, the MTCR coordinates export controls between 37 advanced countries, imposing a "presumption of denial" against heavier platforms like ballistic missiles, while permitting the sale of other technology in a second category which posed less risk for weapons proliferation. These regulations were largely designed with ballistic missiles in mind, but the original 1987 regime also included a provision for unmanned aerial vehicles. As a result, MTCR participants generally refrain from exporting platforms which can travel more than 300 km while carrying a payload more than 500 kg.[59] This category of launch vehicles extends to the more powerful drone models, including the US-made Predator.

On the whole, there is strong adherence to the MTCR. The regime does contain an exemption clause, whereby a country may press ahead with a sale if it judges this to be necessary, but it is only rarely invoked, and has been limited enough to avoid undermining the broader cooperation underlying the MTCR. For instance, the US has tried to amend the wording of the regime in a way which would allow for a more relaxed classification of strategic drones like the Predator, but was denied the necessary consensus vote by partners who remain wary of the fallout that might arise from an unrestrained global market for drones.[60]

China's position on the MTCR is conflicted. The country was not a member of the original 1987 group which produced the regime, but as one of the declared nuclear powers formally recognised by the nuclear Non-Proliferation Treaty (NPT), China is favourably positioned within the global arms control framework, and has a vested interest in supporting the accompanying regimes like the MTCR that strengthen constraints on the spread of weapons of mass destruction. For this reason, despite the country's exclusion from the MTCR, China has repeatedly affirmed that it will unilaterally abide by the terms of the regime. Indeed, Chinese firms maintain that international regulations extend to drone sales, which means that current models marked for export are designed to be controlled by ground stations alone, which limits their operational range, (although it is unclear whether this extends to sub-components like guidance control).[61] Moreover, as one study argues, "the countries importing these UAVs generally lack the necessary command, control, communications, and intelligence capabilities to fully integrate them into effective service".[62]

However, questions remain about China's accountability and transparency over arms deals. During the 1990s, the country was repeatedly sanctioned by the US for violations of the MTCR, but experts are largely in agreement that its official policy is now aligned with international best practice. More recently, there have been several lapses of MTCR-related rule enforcement, as well as close strategic ties with proliferator countries North Korea and Pakistan.[63] Given this chequered background, and uncertainty about the reliability of China's own military transparency in the future, it is unclear whether the lucrative export market will not tempt Chinese industry officials to skirt the country's obligations to arms control.

Another problem flagged by analysts in this area is that unrestrained drone exports may inadvertently flow into the development of chemical, biological, and nuclear weaponry.[64] Even if China is determined not to supply weapons of mass destruction to a particular regime, establishing an arms relationship for unmanned aerial vehicles sales would still provide an instrument well suited for delivery and distribution of these weapons. For example, a loitering drone with a gas tank attached would be a more efficient way to spread toxic agents across a target area.

There is also an overlap between the technology of high-performance drones and cruise missiles, or other delivery systems for nuclear weaponry. This poses challenges for the global effort at strategic arms control: it may well be the case that the sophisticated sub-components in future drone vehicles, such as guidance controls or sensors, are sold individually, and then adapted to missiles. Moreover, as future drones may increasingly resemble the launch vehicles which the MTCR is designed to regulate, the regime is in need of being updated to account for the battlefield potential of lighter drones. This is arguably more concerning in China's case, as the China Haiying Electro-Mechanical Technology Academy, which specialises in aerospace systems and cruise missiles, is also involved in drone research and development; indeed, the Academy was responsible for the jet-powered WJ-600.[65] With the densely interwoven structure of defence firms in China's industrial policy, more of these category-blurring innovations could take place in the future, raising difficult questions for a Chinese foreign policy increasingly mindful of non-proliferation.

Regional Security Concerns

The likely uses of unmanned air power – for expanded ISR, targeted strike in bordering countries, and diplomatic pressure – all have the potential for behaviour that would be concerning to neighbouring countries in the Indo-Pacific. If not handled with care, the presence of unarmed drones conducting shadowing exercises, or pointed harassment of maritime vessels, may trigger a military stand-off or escalating crisis. But so far, China's use of this new military technology has been restrained, which suggests that it is mindful of the complex politics surrounding drones, and the danger of aggravating regional opinion.

There are further steps which could be taken to build on this positive trend. This is all the more compelling because a critical analysis of China's strategic landscape suggests that, for all the access and reach offered by unmanned technology, the country has much to lose from the aggressive use of drones. Instead, the most immediate security implications arising from drones flow from China's security partnerships and defence exports. In and of itself, the introduction of unmanned aircraft by the PLA does not appear likely to cause any dramatic change in the political and security landscape, contrary to some alarmist accounts. If the operational template of drone deployment does change, this would signify a worrying change in the outlook of Chinese foreign policy.

Beyond peace-time deployment, there are several options available for China to indigenously develop drones for military combat in the years ahead. Decoy, anti-submarine, and electronic warfare drones are all seemingly valuable for the PLA, and could be deployed without much alteration to current technology. Other combat concepts and vehicles are more scientifically demanding, and at present, seem unlikely to be realised in any effective way.

This is by no means an exhaustive list of the consequences arising from China's unmanned technology. Indeed, the above pages paint a picture of a Chinese state that is heavily investing in all types of drone vehicles, and poised to exploit unmanned science for many different purposes. Future drones can be expected in ground, water, and sub-surface roles, and may continue to surprise military planners in the years ahead. But it highlights the more likely challenges for regional security that are implied in the PLA's emerging drone capabilities. It is, above, all a reminder that much can be gained by the concerned parties from engaging China on these issues.

Notes

1. Richard Fisher, "Maritime Employment of PLA Unmanned Aerial Vehicles" in Andrew Erickson and Lyle Goldstein, *Chinese Aerospace Power: Evolving Maritime Roles*, 2012.
2. Sandra Erwin, Grace Jean, and Stew Magnuson, "US Troops Vulnerable to Enemy Drones", *National Defense Magazine*, December 2006, accessed at http://www.nationaldefensemagazine.org/archive/2006/December/Pages/WashingtonPulse2798.aspx
3. Maritime drones may be the most consequential in the distant future, but by general consensus among experts, they are the least developed unmanned technology today. See

Samuel Brannen, *Sustaining the Lead in the Unmanned Systems*, Centre for Strategic and International Studies, February 2014 Research Paper, ch. III, n.5, p.11.

4. Lynn Davis et. al., *Armed and Dangerous: UAVs and US Security*, RAND, May 2014 Research Paper, n.4, see ch. 1, n.1, p.9.

5. Sarah Kreps and Micah Zenko, "The Next Drone Wars: Preparing for Proliferation", *Foreign Affairs*, March-April 2012, p.68.

6. Kimberly Hsu et. al., *China's Military Unmanned Aerial Vehicle Industry*, US-China Economic and Security Review Commission, June 2013 Research Report, pp.4-5.

7. Ian Easton and Russel Hsiao, *The Chinese People's Liberation Army's Unmanned Aerial Vehicle Project: Organizational Capabilities and Operational Capabilities Project 2014*, March 2013 Research Report, p.12.

8. Ibid.

9. Ibid., p.14.

10. Tate Krasner, *Assessing China's Unmanned Vehicles: Policy Development, Implementation, Capabilities and Exports*, RSIS Research Manuscript, p.17, accessed at http://careeredge. be.edu/wp-content/uploads/sites/4/2014/08/Chinese UAV/Paper Rewrite.docx, p.17.

11. Paul Scharre, *Robotics on the Battlefield, Part 1: Range, Persistence, and Daring*, Centre for New American Security, May 2014 Report, p.19.

12. Easton and Hsiao, n.7, pp.5-9.

13. Krasner, n.10, p.18.

14. Davis et. al., n.4, p.15.

15. Scott Shane, "Coming Soon: The Drone Arms Race", *New York Times*, October 8, 2011.

16. Shashank Joshi and Aaron Stein, "Emerging Drone Nations," *Survival*, vol.55, n0.5, 2013, p.63.

17. Tabassum Zakaria, "US Drone Policy: Obama Seeking To Influence Global Guidelines", *Reuters*, March 17, 2013.

18. Krasner, n.10, p.6.

19. Austin Strange and Andrews Erickson, "China Has Drones. Now What?," *Foreign Affairs*, May 23, 2013.

20. Brendon Hong, "How China Used Drones to Capture a Notorious Burmese Drug Lord", *The Daily Beast* online, April 17, 2014, accessed at http://www.thedailybeast.com/ articles/2014/04/17/how-china-used-drones-to-capture-a-notorious-burmese-drug-lord. html

21. Wu Jiao and Zhang Yunbi, "Xi Proposes a 'New Silk Road' with Central Asia", *China Daily*, September 8, 2013.

22. Shaun Waterman, "Drone War Heats Up Between China and Japan", *Washington Times*, September 9, 2013.

23. Kristin Roberts, "When the Whole World Has Drones", *National Journal*, March 21, 2013.

24. See, for example, the roundtable, "Is China Losing Faith in North Korea?", *The Guardian*, May 9, 2014.

25. On this point, see Robert Farley, "Should America Fear China's 'Carrier-Killer' Missile?", *The National Interest* online, September 22, 2014, accessed at, http://nationalinterest.org/feature/should-america-fear-chinas-carrier-killer-missile-11321

26. Easton and Hsiao, n.7, p.13.

27. Kreps and Zenko, n.5, p.75.

28. Strange and Erickson, n.19.

29. Greg Torode and Megha Rajagopalan, "Chinese Interceptions of US Military Planes Could Intensify Due to Submarine Base", *Reuters*, August 28, 2014.

30. Easton and Hsiao, n.7, p.5.

31. Peter Singer, *Wired for War: The Robotics Revolution and Conflict in the 21st Century*, 2009, p.56.

32. Defence Science Board, *The Role of Autonomy in DoD Systems*, US Government, July 2012 Repoort, p.74.

33. Fisher, n.1.

34. Easton and Hsiao, n.7, p.14.

35. Defence Science Board, n.32, p.72.

36. See, for example, Mark Gunzinger and Christopher Dougherty, *Outside-In: Operating from Range to Defeat Iran's Anti-Access and Area-Denial Threats*, Centre for Strategic and Budgetary Assessments, January 2012 Research Monograph, pp.47-9.

37. Manjeet Singh Pardesi, "UAVs/UCAVs: Missions, Challenges, and Strategic Implications for Small and Medium Powers," RSIS Working Paper no.66, May 2004, pp.14-15.

38. "Navy Official: China Training for 'Short Sharp War' with Japan", *USNI News*, February 18, 2014.

39. Davis et. al., n.4, p.4.

40. A K Tiwary, "Unmanned Aerial Vehicles in China," *Indian Defence Review*, vol. 28, no.1, January-March, 2013, p.26.

41. Jason Koebler, "Report: Chinese Drone 'Swarms' Designed to Attack American Aircraft Carriers", *US News & World Report*, March 14, 2013.

42. Defence Science Board, n.32, p.72.

43. Roger Cliff et. al., *Shaking the Heavens and Splitting the Earth: Chinese Air Force Employment Concepts in the 21st Century*, RAND, February 2011 Research Monograph, p.189.

44. Easton and Hsiao, n.7, p.5.

45. Ibid., p.10.

46. "Unmanned Systems for EM-Cyber Warfare", navaldrones.com, December 27, 2012, accessed at http://blog.navaldrones.com/2012/12/unmanned-systems-for-em-cyber-warfare.html

47. Krasner, n.10, p.11.

48. Fisher, n.1.

49. Trefor Moss, "Here Come...China's Drone," *The Diplomat* on line, March 2, 2013, accessed http://the diplomat.com/2013/03/herecomes-chinas-drones/?allpoages=yes

50. "Chinese UAVs Will Evolve New Submarine Hunting Methods", unmanned.com, March 1, 2012, accessed at, http://www.unmanned.co.uk/unmanned-vehicles-news/unmanned-aerial-vehicles-uav-news/chinese-uavs-will-evolve-new-subarine-hunting-methods/

51. See, for example, Lyle Goldstein, "Beijing Confronts Long-Standing Weakness in Anti-Submarine Warfare", Jamestown Foundation, *China Brief*, vol.11 no.14, July 29, 2011.

52. See, for example, Thomas Erhard and Robert Work, *Range, Persistence, Stealth, and Networking: The Case for a Carrier-Based Drone*, Centre for Strategic and Budgetary Assessment, 2008 Research Report.

53. Wilson VornDick, "Exploring Unmanned Drones as an Option for China's First Carrier", *Jamestown China Brief*, vol.12, no.7

54. Davis et. al., n.4, p.4.

55. See, for example, Toshi Yoshihara, "Undersea Dragons in the Indian Ocean?", NUS Centre on Asia and Globalisation, China-India Brief no.39, October 14, 2014.

56. Davis et. al., n.4, p.13.

57. Easton and Hsiao, n.7, p.12.

58. Defence Science Board, n.32, p.72.

59. Jefferson Morley, "Drone Proliferation Tests Arms Control", *Arms Control Today*, vol.44, April 2014, accessed at https://www.armscontrol.org/act/2014_04/Drone-Proliferation-Tests-Arms-Control.

60. Micah Zenko and Sarah Kreps, *Limiting Armed Drone Proliferation*, Council on Foreign Relations, Special Report no.69, June 2014, p.18.

61. Krasner, n.10, p.25.

62. Ibid., p.28.

63. See, for example, Ian Stewart and Daniel Salisbury, "Wanted: Karl Lee", *The Diplomat* online, May 22, 2014, accessed at http://thediplomat.com/2014/05/wanted-karl-lee/

64. Joshi and Stein, n.16, p.67.

65. Hsu et. al., n.6, pp.11-12.

CAPS 'Non-Resident Fellowship' Programme on National Security

With a view to reach out to university students, younger defence officers, and professionals (media/academic) interested in research on strategic and defence issues, but not physically based in New Delhi, CAPS has launched a Non-Resident Fellowship Programme focussed broadly on **National Security** issues.

This programme is in keeping with the four core objectives of the Centre:

- Conduct future-oriented, policy-related research on defence and strategic issues to contribute inputs for better understanding of key challenges, their implications, and India's possible responses
- Analyse past, present and future trends in areas of interest to prepare the country as an emerging power in the coming decades.
- Promote a strategic outlook amongst the widest possible populace through publications and seminars
- Spread awareness to stimulate public debate on strategic and security concerns in order to strengthen the country's intellectual capital.

The duration of the fellowship would normally be 9 months and can start at any time of the year. The scholar will be expected to complete a monograph of approximately 30,000 words during the fellowship while working at home/ present location. Applications for fellowship must include a CV and a project proposal (not exceeding 800 words) along with chapterisation. The final manuscript will be reviewed by an independent reviewer for its fitness for publication. If the manuscript is accepted for publication, the Research Fellow will be paid a small honorarium and a certificate from CAPS. For queries and details write to the Centre (e-mail: capsnetdroff@gmail.com) or by letter to following address:

Centre for Air Power Studies
P-284, Arjan Path, Subroto Park, New Delhi 110010
Phone No.: 9125699130,25699132
Fax No.: 911125682533
Email: capsnetdroff@gmail.com

www.ingramcontent.com/pod-product-compliance
Lightning Source LLC
Chambersburg PA
CBHW031441130726
47989CB00003B/1248